W9-CKG-930

New Directions in Applied Mathematics

New Directions in Applied Mathematics

Papers Presented April 25/26, 1980, on the Occasion of the Case Centennial Celebration

Edited by

Peter J. Hilton and Gail S. Young

With Contributions by

Kenneth Baclawski R. W. Brockett
Christopher I. Byrnes Tyrone E. Duncan
Peter J. Hilton D. Kleitman Bertram Kostant
Jerrold E. Marsden Shlomo Sternberg
E. C. Zeeman

Springer-Verlag
New York Heidelberg Berlin

Peter J. Hilton
Department of Mathematics
Case Western Reserve University
Cleveland, OH 44106
U.S.A.

Gail S. Young
Department of Mathematics
University of Wyoming
Laramie, WY 82071
U.S.A.

AMS Classifications: 00A10, 00A69, 90B99, 93B99, 05C99, 58C28, 58C99, 58E99, 93E99, 58F14

Printed in the United States of America.

9 8 7 6 5 4 3 2 1

ISBN 0-387-90604-5 Springer-Verlag New York Heidelberg Berlin
ISBN 3-540-90604-5 Springer-Verlag Berlin Heidelberg New York

Introduction

It is close enough to the end of the century to make a guess as to what the *Encyclopedia Britannica* article on the history of mathematics will report in 2582: "We have said that the dominating theme of the Nineteenth Century was the development and application of the theory of functions of one variable. At the beginning of the Twentieth Century, mathematicians turned optimistically to the study of functions of several variables. But wholly unexpected difficulties were met, new phenomena were discovered, and new fields of mathematics sprung up to study and master them. As a result, except where development of methods from earlier centuries continued, there was a recoil from applications. Most of the best mathematicians of the first two-thirds of the century devoted their efforts entirely to pure mathematics. In the last third, however, the powerful methods devised by then for higher-dimensional problems were turned onto applications, and the tools of applied mathematics were drastically changed. By the end of the century, the temporary overemphasis on pure mathematics was completely gone and the traditional interconnections between pure mathematics and applications restored.

"This century also saw the first primitive beginnings of the electronic calculator, whose development in the next century led to our modern methods of handling mathematics. By the end of the century the symbol-manipulating capabilities of the computer revolutionized mathematics teaching and practice, and the use of the computer to create proofs beyond the capability of the human mind was becoming standard. It was also in the last part of this century that the first large-scale collective proof of a theorem was made, in a classification of objects called finite simple groups.

"Other developments of the century include ..."

I leave the article unfinished both from prudence and from ignorance. I have some trust in my account, but I was glad to hear Michael Atiyah give an elegant talk on the theme of the first two paragraphs at the Karlsruhe International congress on Mathematical Education in 1976 (published in its *Proceedings*).

There are other approaches to this century possible. For example, one could describe this as the century of Hilbert's problems. Certainly these will be mentioned somewhere in the Encyclopedia article. A number of these are concerned with the theme I have proposed for the century. But many are not, and the work originating from some of these will appear much more important in 2582 than I am capable of estimating.

As an example, let me cite Hilbert's Tenth Problem, to construct an algorithm for deciding the existence of integral solutions to polynomials in n variables with integer coefficients. From his statement of the problem, and from other remarks in his paper proposing these problems, it seems probable that he himself had no doubts of the existence of such an algorithm. That such concrete problems turn out to be undecidable is a major fact for pure mathematics. But more followed. One can make a chain from the Tenth Problem to the Turing machine, to von Neumann, to the programmable computer, to IBM, Fortran, my bank statement, and the Four-Color Theorem. The development of the computer cannot be reduced to that sole strand; but that history is over-determined, as psychoanalysts use that term in dream analysis, and this chain is one determination. Future historians may well regard constructivist developments such as the recursive function theory as much more important than many contemporaries now believe, and it may be of great "usefulness." But I would not venture to include it in the first two paragraphs of my Encyclopedia article.

My basic thesis is that the spectacular developments in pure mathematics in the last two decades, coupled with the immense powers of the computer, will revolutionize applications, and soon. Other views are possible. James Frauenthal wrote a note in the April 1980 issue of the *SIAM News* that has been reprinted and that has received much attention. He says, "... we are presently experiencing in mathematics a change which is as dramatic and irreversible as the one which took place in physics earlier in the century. ... The motivating force: the invention of the computer. ... As I see it, within another academic generation, the mainstream of mathematics will not be analysis, number theory and topology, but rather numerical analysis, operations research and statistics. ... By the year 2025 ... in only a few places will there remain centers for research in pure mathematics as we know it today."

Much that I have not quoted I agree with. But it seems clear to me that Frauenthal and others with his view do not understand that there have been reasons for the retreat from applications and that now "algebra, number theory and topology"—and differential topology, logic, algebraic geometry, several complex variables, etc.—are finding applications, and changing not only mathematics but other fields. Who would have thought that physicists would need fiber bundles in quantum mechanics? The computer will make these methods still more powerful and useful. But the computer and its problems will not replace them.

Are there the successes I claim? That of course is the theme of the Conference that provided the papers in this volume, and an answer is simply to say, "Read the book." I had contemplated giving other examples in this introduction, but a recent article by Ian Stewart in Volume 3, No. 2, of the *Mathematical Intellegencer* presents a much better set of illustrations than I possibly could, and I refer the reader to that article.

As the co-directors of the Conference, held at Case Western Reserve University in April, 1980, Peter Hilton and I want to express our gratitude to the speakers. We met with remarkable success in obtaining speakers, and were very pleased with the conference. We hope the reader will agree that something valuable was done.

Our thanks are due to Case Western Reserve University for support of the Conference. We are deeply indebted to Clyde Martin and Marshall Leitman for their invaluable assistance in organizing the Conference, and for suggestions on form and substance.

The University of Wyoming Gail S. Young

Contents

Combinatorics: Trends and Examples

Kenneth Baclawski*

Combinatorics is currently a very active branch of mathematics, and there is good reason to believe that it will become even more active in the future. One trend in combinatorics is the increasing use of methods from other branches of mathematics to solve purely combinatorial questions. A well-known example of this situation is the field of algebraic topology where topological questions are converted into algebraic ones and vice versa, an idea which has been used quite effectively in combinatorics, and which we will illustrate with two examples.

The first example is an interesting chapter in the theory of convex polytopes. This subject has a long history as well as important applications especially to linear and convex programming techniques; however, the "face number problem" that we will be discussing has only recently been solved. The surprising feature of the solution to this problem is that it requires a nontrivial result from the theory of Cohen–Macaulay rings.

Our second example represents a chapter of combinatorics that has only recently begun to be investigated: Discrete Fixed Point Theory. Yet some remarkable and unexpected results are already known that lead one to suspect that a beautiful theory lies waiting to be developed.

The author would like to express his gratitude to Gian-Carlo Rota. He has been a constant source of ideas and inspiration.

Face Numbers of Convex Polytopes

A (*bounded*) *convex polytope* is a bounded subset of Euclidean n-space $\mathbb{R}^n$, defined by linear equations and inequalities. The boundary of such a figure

* Haverford College, Haverford, PA 19041.

breaks up into *faces* of various dimensions. The highest dimensional ones are called *facets*. Let us call d the dimension of the facets. A *vertex* (or *extreme point*) is a face of dimension 0, an *edge* is a face of dimension 1, and so on. Let f_k be the number of faces of dimension k. We call f_k the kth *face number* of the polytope, and the $(d+1)$-tuple $(f_0, f_1, \ldots, f_d)$ is called the *face vector*. For example, the cube in $\mathbb{R}^3$ has 8 vertices, 12 edges, and 6 facets, so that its face vector is

$$(f_0, f_1, f_2) = (8, 12, 6)$$

If the interior of the polytope is nonempty, then the dth face number f_d is the number of nonredundant "constraints" that are required to define the polytope. In linear programming one is given a set of constraints. Thus a natural "complexity" question is to find an upper bound for each face number f_i, given that f_d is known. An even more ambitious problem is to characterize the possible $(d+1)$-tuples that can occur as face vectors. Such questions are interesting both because of the intrinsic beauty of the subject and because of applications to the question of the complexity of linear programming and related problems.

The most important class of convex polytopes is the set of simple polytopes: a polytope is *simple* if every face of dimension k is contained in exactly $d - k + 1$ facets. Roughly speaking, simple polytopes result from "random" or "generic" constraints. More precisely, given any convex polytope the facets can be perturbed slightly so as to convert the polytope to a simple one. In this process, the number of facets remains unchanged while the other face numbers get no smaller. For a proof of this see McMullen–Shephard (1971). For example, the square pyramid is not simple, but a slight perturbation of two of the "side facets" will produce a simple polytope.

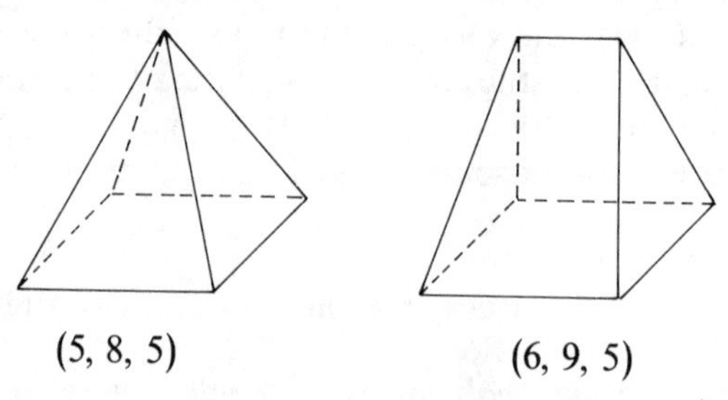

face vectors: (5, 8, 5) (6, 9, 5)

The first step in the study of the face vectors of simple polytopes is to construct the *difference triangle* as follows. Write the face numbers in reverse order down a diagonal, and write a sequence of ones along the opposite diagonal:

$$\begin{array}{ccccccccc}
 & & & & 1 & & & & \\
 & & & 1 & & f_d & & & \\
 & & 1 & & & & f_{d-1} & & \\
 & \cdot^{\cdot^{\cdot}} & & & & & & \ddots & \\
 & 1 & & & & & & & f_0 \\
1 & & & & & & & &
\end{array}
\qquad\qquad
\begin{array}{ccccccc}
 & & & 1 & & & \\
 & & 1 & & 6 & & \\
 & 1 & & & & 12 & \\
1 & & & & & & 8 \\
\end{array}$$

$$\begin{array}{ccc}
\text{general case} & \qquad\qquad & \text{the cube}
\end{array}$$

Next below each pair of integers in the triangle, enter the difference of that pair, the right number minus the left one. For the cube the completed triangle looks like this:

$$\begin{array}{ccccccccc}
 & & & & 1 & & & & \\
 & & & 1 & & 6 & & & \\
 & & 1 & & 5 & & 12 & & \\
 & 1 & & 4 & & 7 & & 8 & \\
1 & & 3 & & 3 & & 1 & &
\end{array}$$

The bottom row of the difference triangle is called the *h-vector* of the polytope. In general the h-vector is a $(d+2)$-tuple, $(h_0, h_1, \ldots, h_{d+1})$, with $h_0 = 1$ and $h_1 = f_d - d - 1$. It is an easy exercise to show that the face vector is determined by the h-vector; indeed, each face number is a linear combination of the h-numbers, with non-negative integral coefficients. Thus upper bounds on the h-numbers yield upper bounds on the face numbers, but not vice versa, and characterizing the face vectors is equivalent to characterizing the possible h-vectors.

The first feature of the h-vector for the cube that one notices is its symmetry: it reads the same in either direction. This is no accident and is equivalent to the *Dehn–Sommerville equations.* One proves that the h-vector is symmetrical for a simple polytope by using an Euler characteristic argument. See McMullen–Shephard (1971). However, the Dehn–Sommerville equations are not sufficient to characterize the possible h-vectors. The h-vectors also have the property that the h-numbers do not increase "too rapidly."

The first example of such a property satisfied by the h-numbers is the Upper Bound Conjecture made by Motzkin (1957) and proved by McMullen (1971). It states that

$$0 \leq h_k \leq \binom{h_1 + k - 1}{k} = \binom{f_d + k - d - 2}{k}.$$

McMullen proved this conjecture by arranging the vertices in an order convenient for counting the faces of each dimension. Such an ordering on the vertices is called a *shelling.* That simple polytopes can be shelled was shown by Bruggesser and Mani (1971).

McMullen went on to conjecture that the h-vector satisfies an even stronger property. To state this property we need some notation. It is an easy exercise to show that, for any non-negative integer a and positive integer k, there are unique integers $j, m_j, \ldots, m_k$ satisfying $m_k > m_{k-1} > \cdots > m_j \geq j \geq 1$ such that

$$a = \binom{m_k}{k} + \binom{m_{k-1}}{k-1} + \cdots + \binom{m_j}{j}.$$

A sequence $a_1, a_2, \ldots,$ is said to be an *M-sequence* if for any $k \geq 1$ we have

$$0 \leq a_{k+1} \leq \binom{m_k + 1}{k+1} + \binom{m_{k-1} + 1}{k} + \cdots + \binom{m_j + 1}{j+1},$$

where $j, m_j, \ldots, m_k$ are chosen as above for $a = a_k$. McMullen conjectured that the h-vector of a convex polytope is an M-sequence. Although McMullen made this conjecture, as he proved the Upper Bound Conjecture, by using a keen geometric insight into the problem, the proof of his conjecture involved concepts quite far removed from geometry.

Before discussing the proof of McMullen's conjecture, we digress a little to a seemingly unrelated topic. The concept of an M-sequence is named not after McMullen but after Macaulay (1927) who showed the following result. Let S be a commutative ring obtained from a polynomial ring by setting a collection of homogeneous polynomials equal to zero:

$$S = \mathbb{Q}[X_1, \ldots, X_n]/(p_1(X), \ldots, p_m(X)).$$

Here the "$\mathbb{Q}$" stands for the field of rational numbers, and $\mathbb{Q}[X_1, \ldots, X_n]$ is the ring of polynomials in the variables $X_1, \ldots, X_n$ with rational coefficients. Since the polynomials $p_1(X) = p_1(X_1, \ldots, X_m), \ldots, p_m(X)$ are homogeneous, it makes sense to speak of the *degree* of the class of a homogeneous polynomial in S. Macaulay showed that if g_k is the dimension (as a rational vector space) of the space of polynomial classes in S of degree k, then $g_1, g_2, \ldots,$ is an M-sequence. More precisely,

Macaulay's Theorem. *If $p_1(X_1, \ldots, X_n), \ldots, p_m(X_1, \ldots, X_n)$ are homogeneous polynomials and if S is the quotient ring*

$$S = \mathbb{Q}[X_1, \ldots, X_n]/(p_1(X), \ldots, p_m(X)),$$

then the sequence of dimensions $\dim(S_1), \dim(S_2), \ldots$ *is an M-sequence, where S_k is the homogeneous part of S of degree k.*

Macaulay proved his theorem by first reducing to the special case for which the polynomials $p_1(X), \ldots, p_m(X)$ are monomials. The result in this case is "purely combinatorial." To prove it he used a "compression" technique. The independent monomials in S of a given degree are replaced by the same number of monomials occurring first in a suitable lexicographic sense. The binomial coefficients in the definition of an M-sequence count the number of monomials of a specific kind after compression. This technique has been generalized by Clements and Lindström (1969) who also obtain the Kruskal (1963)–Katona (1966) theorem as a special case, and by Metropolis–Rota (1978) who consider the cubical analogue. For a good treatment of this topic see Greene–Kleitman (1978).

The fact that the concept of an M-sequence occurs both in Macaulay's Theorem and in McMullen's conjecture is strongly suggestive of a connection between the two. This connection was established by Stanley (1975) who showed the following. Let P be a simple polytope with set of facets $\mathscr{F}$. For each facet $F \in \mathscr{F}$, let $X(F)$ be an indeterminate ("variable"). Define a ring R as the quotient ring obtained from the polynomial ring $\mathbb{Q}[X(F) | F \in \mathscr{F}]$ by setting a monomial $X(F_1)X(F_2) \cdots X(F_l)$ equal to zero whenever $F_1 \cap F_2 \cap \cdots \cap F_l = \varnothing$. Clearly the ring R carries combinatorial information about the polytope. Indeed, a monomial $X(F_1)X(F_2) \cdots X(F_l)$ is nonzero in R if and only if $F_1 \cap F_2 \cap \cdots \cap F_l$ is a face of P (the F_i's need not be distinct), and these monomials form a basis of R.

To relate the ring R to the face vector, Stanley used a ring-theoretical concept introduced by Macaulay. A ring S (obtained as above) is said to be *Cohen–Macaulay* if there are linear homogeneous polynomials $q_1(X), \ldots, q_d(X)$ such that

(1) for every k, $q_k(X)$ is not a zero-divisor in $S/(q_1(X), \ldots, q_{k-1}(X))$;

(2) $S/(q_1(X), \ldots, q_d(X))$ is finite dimensional as a rational vector space.

Stanley then showed that if his ring R is Cohen–Macaulay then the McMullen conjecture is true. More precisely,

Stanley's Theorem. *Let P be a simple polytope, and let R be the ring defined above. If $q_1(X), \ldots, q_d(X)$ satisfy* (1) *and* (2) *above, then the* k*th* h*-number is the dimension of the homogeneous part of degree k of*

$$R/(q_1(X), \ldots, q_d(X)).$$

Thus Stanley converted McMullen's conjecture into a problem in commutative algebra.

It is an amazing coincidence that a commutative algebraist, Reisner (1976), working independently of Stanley, showed that Stanley's ring R is Cohen–Macaulay. Indeed Reisner characterized precisely when a ring, obtained by setting square-free monomials equal to zero, is Cohen–Macaulay. His proof made use of some elaborate machinery from commutative algebra; and although much simpler proofs are now known (cf. Baclawski–Garsia (1981)), Reisner's result was quite an achievement. It is all the more surprising because he was motivated by no particular application. His was a "purely mathematical" investigation, yet it has more applications than virtually any other recent result in commutative algebra. A situation like this makes one wonder whether the commonly accepted distinction between pure and applied mathematics is valid.

The theory of face vectors did not end with the Stanley–Reisner result. McMullen had actually gone on to conjecture an even stronger property of the h-vector which he felt would completely characterize it. To state this we return to the difference triangle. By the Dehn–Sommerville equations, the first half of the h-vector determines everything. Using this first half, continue the difference triangle one more line. McMullen conjectured that this last line is also an M-sequence and that this property characterizes the h-vectors

$$
\begin{array}{ccccccccccc}
 & & & & & 1 & & & & & \\
 & & & & 1 & & 6 & & & & \\
 & & & 1 & & 5 & & 12 & & & \\
 & & 1 & & 4 & & 7 & & 8 & & \\
 & 1 & & 3 & & 3 & & 1 & & & \\
1 & & 2 & & & & & & & &
\end{array}
$$

extended difference triangle
for the cube

and hence the face vectors of convex polytopes. This conjecture was proved recently by Stanley (1980) and Billera–Lee (1980). The proof involves the "hard" Lefschetz Theorem as well as the Stanley–Reisner ring.

Stanley made a discovery of a connection between a combinatorial problem in the theory of convex polytopes and a property of commutative rings. This connection has had many other applications since then, and it clearly represents mathematics at its best, whether one chooses to call it pure or applied. How does one prepare a student to do mathematics of this kind? Or more generally, how should one teach mathematics? Clearly a broad background is important. Stanley had to know enough Commutative Algebra to have been aware of Macaulay's work and to be able to take advantage of it. On the other hand one must also convince students to break out of established methods and to risk trying something new. An important aspect of teaching should therefore be to free students from their preconceptions.

Discrete Fixed Point Theory

Our second example is of quite a different character from the first. Whereas the first is essentially complete, the second is only beginning: it is a theory in its infancy. Nevertheless some results are already known that are quite remarkable and unexpected.

Discrete fixed point theory deals with the following situation. We have a partially ordered set P (poset) with order relation "$\leq$" and a mapping $f: P \to P$ such that f preserves the ordering, i.e., $x \leq y$ implies that $f(x) \leq f(y)$. The *fixed point set* of f is the subset $P^f = \{x \in P \mid f(x) = x\}$. We would like to know what kind of properties P^f has. The simplest property is whether or not P^f is nonempty. We say that P has the *fixed point property* if P^f is nonempty for every f.

There are a number of situations where discrete fixed point theory arises naturally. For example, let M be a compact manifold and $g: M \to M$ a continuous map. This is the situation we have in topological fixed point theory. Assume that M can be triangulated by a triangulation Δ such that g can be approximated by a simplicial map $f: \Delta \to \Delta$. Now a triangulation has the structure of a poset: the elements are the simplices and the ordering is containment. Moreover $f: \Delta \to \Delta$ is an order-preserving map. The fixed point set Δ^f is an approximation to the fixed point set $M^g = \{p \in M \mid g(p) = p\}$.

In fact, discrete fixed point theory can be applied to more general "decompositions" of a manifold than just triangulations. The "cubical complexes" of Metropolis–Rota (1978) may also be used. There are also interesting connections with game theory and the theory of representations of finite groups. See Baclawski–Björner (1981b) for discussion and examples.

For another example, let B be a Banach space and let $T: B \to B$ be a continuous (bounded) linear operator. By the bounded inverse theorem of functional analysis, if T is a bijection then T^{-1} is also a continuous linear operator and hence T is an automorphism of B. Let $P(B)$ be the poset of *proper* subspaces of B, i.e., $P(B) = \{0 \subsetneqq V \subsetneqq B \mid V \text{ is a subspace}\}$. Let $P_0(B) \subseteq P(B)$ be the subposet of closed proper subspaces of B. Then T induces order-preserving automorphisms of $P(B)$ and $P_0(B)$. The fixed point sets $P(B)^T$ and $P_0(B)^T$ are the sets of *invariant subspaces*, resp., closed invariant subspaces, of the operator T and are of much current interest in functional analysis.

The oldest result in discrete fixed point theory is the Tarski (1955)–Davis (1955) Theorem. This theorem characterizes the fixed point property for lattices. A poset P is said to be a *lattice* if every finite subset of P has both a least upper bound and a greatest lower bound. A lattice is *complete* if *every* subset has these. The Tarski–Davis Theorem states that a lattice has the fixed point property if and only if it is complete, and in this case the fixed point sets are also complete lattices.

Fixed point theory for posets in general has turned out to be more complicated than the Tarski–Davis Theorem suggests. So far the only fixed point theorem we have for posets in general applies only to finite posets and is the analogue for posets of the Lefschetz Fixed Point Theorem of algebraic topology. This result was proved by Baclawski–Björner (1979) and is called the *Hopf–Lefschetz Fixed Point Theorem.* To state it we need some notation from algebraic topology.

Let P be a finite poset. Write $\Delta(P)$ for the simplicial complex whose vertices are the elements of P and whose simplices are the totally ordered subsets (*chains*) of P: $\{x_1 < x_2 < \cdots < x_n\} \subseteq P$ is a typical simplex. Let $|\Delta(P)|$ be the polyhedron defined by the simplicial complex $\Delta(P)$. We write

$\tilde{H}_i(P, \mathbb{Q})$ for the reduced rational homology (in dimension i) of the space $|\Delta(P)|$, i.e.,

$$\tilde{H}_i(P, \mathbb{Q}) = \tilde{H}_i(|\Delta(P)|; \mathbb{Q}).$$

The *reduced Euler characteristic* of P is defined to be

$$\mu(P) = \sum_{i=-1}^{\infty} (-1)^i \dim_{\mathbb{Q}} \tilde{H}_i(P, \mathbb{Q}) = \chi(|\Delta(P)|) - 1,$$

where $\chi(X)$ is the usual Euler characteristic of a space X.

Now suppose that $f: P \to P$ is an order-preserving map. Then f induces a simplicial map $\Delta(f): \Delta(P) \to \Delta(P)$ and a continuous map $|f|: |\Delta(P)| \to |\Delta(P)|$. It then follows that f induces linear transformations

$$\tilde{f}_i: \tilde{H}_i(P, \mathbb{Q}) \to \tilde{H}_i(P, \mathbb{Q}).$$

Since each $\tilde{f}_i$ is a linear transformation from a vector space to itself, it makes sense to speak of the trace of $\tilde{f}_i$. The *Lefschetz number* of f is the alternating sum

$$\Lambda(f) = \sum_{i=-1}^{\infty} (-1)^i \operatorname{Trace}(\tilde{f}_i).$$

We then have:

Hopf–Lefschetz Fixed Point Theorem. *If P is a finite poset and if $f: P \to P$ is an order-preserving map, then*

$$\Lambda(f) = \mu(P^f).$$

The most important special case of the Hopf–Lefschetz Theorem is the following. A finite poset P is said to be *acyclic* if $\tilde{H}_i(P, \mathbb{Q}) = 0$ for every i. For example, if P consists of only one element or if there is an element $x \in P$ such that x is comparable with every other element of P, then P is acyclic. The empty poset is not acyclic because $\tilde{H}_{-1}(\varnothing, \mathbb{Q}) = \mathbb{Q}$.

Corollary. *If P is a finite acyclic poset, then P has the fixed point property.*

To prove the corollary just note that $\Lambda(f) = 0$ for any order-preserving map $f: P \to P$. Hence $\mu(P^f) = 0$ for any such map. Since $\mu(\varnothing) = -1$, we must have that $P^f \neq \varnothing$. We remark that the topological analogue of the corollary above is true but that the analogue of the Hopf–Lefschetz Theorem is not. Most applications of the Hopf–Lefschetz Theorem use only the corollary.

Of course the hard part is to show that a given poset is acyclic. This can be quite difficult. However, some very remarkable results have been shown; and, like the work of Reisner, they were often the result of purely theoretical investigations with no immediate applications to motivate them. An example of this was found by the author (cf. Baclawski (1977)) who thereby uncovered a hitherto unsuspected connection between fixed points and complements in a lattice.

Let L be a finite lattice. Write $\bar{L}$ for the proper part of the lattice, i.e., $\bar{L} = L - \{0, 1\}$. Elements x and y of L are said to be *complements*, and we write $x \perp y$, in this case, if $x \vee y = 1$ and $x \wedge y = 0$. A lattice is said to be *complemented* if every element of the lattice has a complement, and the lattice is *noncomplemented* otherwise. In Baclawski–Björner (1981a) we find these two results:

(1) If an order-preserving map $f\colon \bar{L} \to \bar{L}$ has no fixed points, then L is complemented.
(2) If $f\colon L \to L$ is an order-preserving automorphism and if $x \in \bar{L}$ has the property that $x \wedge z = 0$ for every $z \in P^f$, then there is an element $y \in \bar{L}$ such that $y \perp x$ and $f^n(y) \wedge x \neq 0$ for some integer n.

We think of f as defining a discrete dynamical system where n represents the time, in which case (2) represents a weak "mixing" property. Think of elements x, y as being "close" if $x \wedge y \neq 0$ and as being "far apart" if $x \perp y$. Statement (2) then tells us that if x isn't close to any of the fixed points of f then the orbit of x comes close to some element far from x. Those interested in this result and in discrete fixed point theory in general should look at the forthcoming survey article by Baclawski–Björner (1981b).

Both statements (1) and (2) are proved by showing that a certain poset is acyclic. For example, statement (1) says that if L is noncomplemented then $\bar{L}$ has the fixed point property. This follows from a result of Baclawski (1977) which states that if $x \in L$ has no complements then $\bar{L}$ is acyclic. The same proof used there has now been improved to give the following result of Baclawski–Björner (1981a) from which statement (2) is derived.

Theorem. *Let L be a finite lattice and let $x \in \bar{L}$. Suppose that B satisfies*

$$\{y \in \bar{L} \mid y \perp x\} \subseteq B \subseteq \{z \in \bar{L} \mid z \wedge x = 0\}.$$

Then $\bar{L} - B$ is acyclic.

So far no purely combinatorial proofs are known for statements (1) and (2) above, and it is not yet clear how to generalize the results to infinite posets. As a result these theorems have no direct application to posets such as $P(B)$ or $P_0(B)$ for a Banach space B. Nevertheless, by uncovering the relationship between fixed points and complements in lattices, the Baclawski–Björner theorems lead one to suspect a similar relationship in other settings.

This example illustrates an interesting point about the complexity of the interaction between a mathematical theory and its applications. Namely, a theorem can result not in an answer or solution to a problem in some other area, but rather can suggest new *problems* worth being studied. This reverses the conventional role that mathematics plays in applications in which mathematics is asked to solve or at least shed some light on problems posed in some other discipline. In this same volume Professor Zeeman has given excellent examples where the roles of " theory " and " applications " are more balanced. Physical experiments can be used to suggest the direction of future research in catastrophe theory and a model in catastrophe theory has led to an interesting medical research problem concerning pituitary function that had never before been considered.

References

1. K. Baclawski, Galois connections and the Leray spectral sequence, *Advances in Math.* **25** (1977), 191–215.
2. K. Baclawski and A. Björner, Fixed points in partially ordered sets, *Advances in Math,* **31** (1979), 263–287.
3. ——, Fixed points and complements in finite lattices, *J. Comb Theory, Ser. A,* to appear, 1981.
4. ——, Fixed points in ordered structures: a survey, in preparation, 1981.
5. K. Baclawski and A. Garsia, Combinatorial decompositions of a class of rings, *Advances in Math.,* **39** (1981), 155–184.
6. L. Billera and C. Lee, Sufficiency of McMullen's conditions for f-vectors of simplicial polytopes, *Bull. Amer. Math. Soc.* (N.S.), **2** (1980), 181–185.
7. H. Bruggesser and P. Mani, Shellable decompositions of cells and spheres, *Math. Scand.* **29** (1971), 197–205.
8. G. Clements and B. Lindström, A generalization of a combinatorial theorem of Macaulay, *J. Comb. Theory* **7** (1969), 230–238.
9. A. Davis, A characterization of complete lattices, *Pacific J. Math.* **5** (1955), 311–319.
10. C. Greene and D. Kleitman, Proof techniques in the theory of finite sets, in *Studies in Combinatorics, M.A.A. Studies in Math.,* vol. 17, G.-C. Rota, ed.; Math. Assoc. of Amer., Providence, RI, 1978, 22–79.
11. G. Katona, A theorem of finite sets, in *Proc. Tihany Conf., 1966,* Budapest, 1968.
12. J. Kruskal, The number of simplices in a complex, in *Mathematical Optimization Techniques,* Univ. California Press, Berkeley, 1963, 251–278.
13. F. Macaulay, Some properties of enumeration in the theory of modular systems, *Proc. London Math. Soc.* **26** (1927), 531–555.
14. P. McMullen, The number of faces of simplicial polytopes, *Israel J. Math.* **9** (1971), 559–570.
15. P. McMullen and G. C. Shephard, *Convex Polytopes and the Upper Bound Conjecture,* London Math. Soc. Lect. Note Ser. 3, Cambridge Univ. Press, 1971.
16. N. Metropolis and G.-C. Rota, Combinatorial structure of the faces of the n-cube, *SIAM J. Appl. Math.* **35** (1978), 689–694.
17. T. Motzkin, Comonotone curves and polyhedra, Abstract 111, *Bull. Amer. Math. Soc.* **63** (1957), 35.
18. G. Reisner, Cohen–Macaulay quotients of polynomial rings, *Advances in Math.* **21** (1976), 30–49.
19. R. Stanley, The upper bound conjecture and Cohen–Macaulay rings, *Stud. in Appl. Math.* **54** (1975), 135–142.
20. —, The number of faces of a simplicial convex polytope, preprint, M.I.T., 1980.
21. A. Tarski, A lattice-theoretical fixpoint theorem and its applications, *Pacific J. Math.* **5** (1955), 285–309.
22. C. Zeeman, Bifurcations, catastrophes and turbulence, this volume, 109–153.

Control Theory and Singular Riemannian Geometry*

R. W. Brockett†

This paper discusses the qualitative and quantitative aspects of the solution of a class of optimal control problems, together with related questions concerning a corresponding stochastic differential equation. The class has been chosen to reveal what one may expect for the structure of the set of conjugate points for smooth problems in which existence of optimal trajectories is not an issue but for which Lie bracketing is necessary to reveal the reachable set. It is, perhaps, not too surprising that in thinking about this problem various geometrical analogies are useful and, in the final analysis, provide a convenient language to express the results. Indeed, the geodesic problem of Riemannian geometry is commonly taken to be the paradigm in the calculus of variations; a point of view which is supported by a variety of variational principles such as the theorem of Euler which identifies the path of a freely moving particle on a manifold with a geodesic and the whole theory of general relativity. Nonetheless, the class of variational problems considered here can only be thought of as geodesic problems in some limiting sense in which the metric tends to infinity. For this reason the geodesic analogy has to be developed rather carefully. What is actually needed is a generalization of Riemannian geometry and it seems that the intuitive content of Riemannian geometry is sufficiently robust so as to withstand modifications of the type required and still provide a reasonably "geometric" picture. We consider questions involving model spaces, geodesic equations, the appropriate definition of the Laplace–Beltrami operator, etc. The end results make avail-

* This research was supported in part by the Army Research Office under Grant DAAG29-76-C-0139, the U.S. Office of Naval Research under the Joint Services Electronics Program Contract N00014-75-C-0648 and the National Science Foundation under Grant ENG-79-09459 at the Division of Applied Sciences, Harvard University, Cambridge, MA.

† Division of Applied Sciences, Harvard University, Cambridge, MA 02138.

able in the control setting, considerable geometrical insight and suggest some novel problems in differential geometry.

In addition to work in control theory and geometry which we draw on in a very specific way, one sees in the recent work of physicists an exciting, albeit vague, parallelism centering around the idea of "superspace." Before embarking on the actual mathematics of this paper let me make a few comments on this. Part and parcel of the Riemannian expression for infinitesimal distance

$$(ds)^2 = \Sigma g_{ij}(x)\, dx_i\, dx_j$$

is idea that space is "essentially isotropic." That is to say, the distance to nearby points involves the same kind of expression regardless of the direction. Characteristic of the models which we investigate here is a very strong anisotropic character as would be suggested by an expression such as $(ds)^2 = (dx)^2 + (dy)^2 + |dz|$. There have been, and continue to be, suggestions in the physics literature to the effect that what we perceive as being a four dimensional space–time continuum may be better thought of as being a submanifold of a higher dimensional space. In the theory of O. Klein and T. Kaluza (see [1]) one takes the ambient space to be five dimensional, obtaining in return a setting in which electromagnetic and gravitational theories are unified. In more recent work, e.g., Zumino's article in [2], one sees suggestions about ten and twenty six dimensional ambient spaces. Manifestly these theories refer to a highly anisotropic kind of space. Having planted the idea that what is to be discussed here may have physical as well as mathematical interest we hasten to add that only the mathematical and control theoretic aspects will be considered further.

Optimal control and geodesics have been discussed before in the literature, for example Hermes [3] and Hermann [4], however the most directly relevant prior work that I am aware of occurs in the thesis of J. Baillieul [5] where he carries out certain detailed computations on a specific model of the type considered here.

I thank the organizers of the conference for giving me the opportunity to speak at my alma mater on the occasion of its 100th anniversary. It was a pleasant occasion. I also want to express my appreciation to J. Baillieul, C. I. Byrnes and N. Gunther for their patience in listening to, and help in clarifying the arguments given here.

The Starting Point

Consider a neighborhood of x_0 in n-dimensional Cartesian space $\mathbb{R}^n$, and consider the following problem from control theory. Given

$$\dot{x} = B(x)u, \qquad \dot{x} = \frac{d}{dt}x$$

find $u(t) \in \mathbb{R}^m$ on the interval $[0, 1]$ such that $x(0) = x_0$, $x(1) = x_1$, and

$$\eta(x_0, x_1) = \int_0^1 (\langle u, u \rangle)^{1/2}\, dt$$

is minimized. Here $\langle\ ,\ \rangle$ denotes the standard inner product on $\mathbb{R}^m$. We investigate this problem under the assumption that B is smooth and of constant rank m. In place of η we study

$$\rho(x, y) = \min_u \eta(x, y).$$

Notice that ρ satisfies the condition $\rho(x, x) = 0$, $\rho(x, y) = \dot{\rho}(y, x) > 0$ if $x \neq y$ and $\rho(x, y) \leq \rho(x, z) + \rho(z, y)$. That is, ρ satisfies the axioms of a metric. The only step here which is not completely obvious is $\rho(x, y) = \rho(y, x)$ and this is proven by replacing $u_i(t)$ by $-u_i(1 - t)$ and noticing that this control steers y to x if u steers x to y.

In the special case where $m = n$, under our announced hypotheses we may rewrite $\dot{x} = B(x)u$ as $B^{-1}(x)\dot{x} = u$ and express the problem as a Riemannian geodesic problem, i.e., to find from among all smooth paths joining x and y the one which minimizes

$$\eta = \int_0^1 (\langle B^{-1}(x)\dot{x}, B^{-1}(x)\dot{x} \rangle)^{1/2}\, dt.$$

Thus we see that $(B^{-1}(x))^T B^{-1}(x) = G(x)$ plays the role of the metric tensor if $B^{-1}(x)$ exists. However, $\rho(x, y)$ may be well defined even if B is not invertible and in particular even if $m < n$. All that is needed for $\rho(x, y)$ to be defined is that every point should be reachable from every other point. None of the phenomena which we investigate are a consequence of any lack of smoothness in B or the quantity being minimized; for the sake of simplicity we take B to be C^∞ although we could get by with less.

What are the conditions for every point near x to be reachable from x? This kind of question is studied in the control literature under the names controllability, reachability, etc. but the specific result we need was known already by Chow [6] who generalized a result of Caratheodory. What is needed is that the Lie algebra of vector fields generated by

$$F_i = \sum_{i=1} b_j^i \frac{\partial}{\partial x^j}, \qquad B = (b_j^i)$$

should be sufficiently rich to span $\mathbb{R}^n$ at each point. This condition is considerably less demanding than the condition that B is invertible!

Perhaps an example will be of some help in developing intuition. Consider the following prototype for the situation in $\mathbb{R}^3$:

$$\begin{aligned} \dot{x} &= u \\ \dot{y} &= v \\ \dot{z} &= uy - vx. \end{aligned}$$

In this case

$$F_u = \frac{\partial}{\partial x} + y\frac{\partial}{\partial z}; \qquad F_v = \frac{\partial}{\partial y} - x\frac{\partial}{\partial z}, \qquad [F_u, F_v] = 2\frac{\partial}{\partial z}.$$

Since these span $\mathbb{R}^3$ we can reach any point from any other point. However, B is 3 by 2 and so $B^T B$ is not invertible and we are not in the standard Riemannian situation. With the help of the Lagrange multiplier technique one can show that the geodesics satisfy

$$\ddot{x} + \lambda\dot{y} = 0$$
$$\ddot{y} - \lambda\dot{x} = 0$$
$$\ddot{z} + \lambda(\dot{x}x + \dot{y}y) = 0$$

where λ is a suitable constant. In fact, from the last equation we see that for trajectories which pass through (0, 0, z) we have

$$\lambda = \frac{\dot{z}}{x^2 + y^2}.$$

The locus of points equidistant from (0, 0, 0) displays an x_3-axis symmetry but, in contrast with the Riemannian situation, the geodesic spheres are not smooth manifolds. (They fail to be smooth at the north and south poles.)

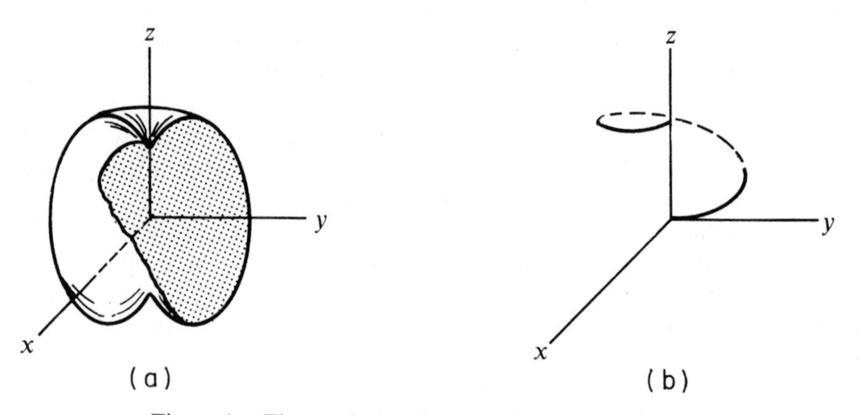

Figure 1 The geodesic spheres and one geodesic curve.

We can think of this example in the following way. At each point in the space we have a two dimensional subspace of the tangent space, the one spanned by the vector fields

$$\frac{\partial}{\partial x} + y\frac{\partial}{\partial z} \quad \text{and} \quad \frac{\partial}{\partial y} - x\frac{\partial}{\partial z}.$$

In this plane we have a given inner product corresponding to the fact that we are minimizing the integral of $u^2 + v^2$. We may think of this plane as being a two space of *ordinary directions*. In this problem the geodesics emanating from a point are characterized by an initial velocity chosen from the ordinary directions together with parameter λ which controls, in a way we want to make precise, the amount of twist the trajectory has to bring it away from the plane of ordinary directions.

It may also be pointed out that for this example the points conjugate to the point (0, 0, 0) consist of the entire z-axis. Recall [7] that in an ordinary Riemannian space the points conjugate to p have distance

$$\rho(p, q) \geq \frac{\pi}{\sqrt{K}}$$

where K is the maximum sectional curvature of the manifold. Since p is conjugate to points in every neighborhood of it we see that we are dealing with a space having rather exceptional curvature!

Naturally associated with this problem is a subgroup of the affine group on $\mathbb{R}^3$ consisting of elements of the form

$$\begin{bmatrix} x \\ y \\ z \end{bmatrix} \mapsto \begin{bmatrix} \theta_{11} & \theta_{12} & 0 \\ \theta_{21} & \theta_{22} & 0 \\ \alpha & \beta & 1 \end{bmatrix} \begin{bmatrix} x \\ y \\ z \end{bmatrix} + \begin{bmatrix} -\beta \\ +\alpha \\ \gamma \end{bmatrix}, \qquad \begin{bmatrix} \theta_{11} & \theta_{12} \\ \theta_{21} & \theta_{22} \end{bmatrix} = \text{orthogonal}.$$

This group acts transitively on $\mathbb{R}^3$ and leaves the form of the variational problem invariant. Thus the calculation of $\rho(\cdot, \cdot)$ is no more difficult than the calculation of $\rho(0, \cdot)$. Based on this remark we can see that just as through (0, 0, 0) there is a line of points $\{p \mid p = (0, 0, z)\}$ which are conjugate to (0, 0, 0), there is a line of points $\{p \mid p = (a, b, z)\}$ which are conjugate to (a, b, z_0). At each point in $\mathbb{R}^3$ this gives us a natural splitting of the tangent space into a two dimensional subspace Range B and a one dimensional subspace defined by tangent vector to the manifold of conjugate points.

Finally, there is a second order operator associated with this problem, namely

$$L = \left(\frac{\partial}{\partial x} + y\frac{\partial}{\partial z}\right)^2 + \left(\frac{\partial}{\partial y} - x\frac{\partial}{\partial z}\right)^2,$$

which shares many of the properties of the heat operator. We will discuss this further in the final section of the paper.

The Hamiltonian Formulation

We now return to the general situation and set about the problem of studying the geodesics. It saves a certain amount of annoying calculation to observe right at the start that the trajectories which minimize η also minimize

$$\tilde{\eta} = \int_0^1 \langle u, u \rangle \, dt.$$

This comes about, as it does in the case of Riemannian geometry, because the value of $\langle u, u \rangle$ along geodesic curves is constant.

In Riemannian geometry the equations for the geodesics can be written as equations on the tangent bundle. Choosing coordinates, these may be expressed in terms of the Levi–Civita connection as

$$\ddot{x}^i + \Gamma^i_{jk} \dot{x}^j \dot{x}^k = 0 \quad \text{(summation convention)}.$$

In the present situation the tangent bundle formulation is not quite so straightforward. Instead, we begin with a Hamiltonian formulation on the cotangent bundle. According to the maximum principle of optimal control Hamilton–Jacobi theory in the present context we may associate with the geodesic problem a pair of first order equations

$$\dot{x} = Bu$$

$$\dot{p} = A(u, p),$$

where A is a bilinear form in u and p, and assert that if $x(\cdot)$ is a geodesic then there exists a $p(0)$ such that (x, p) satisfy these equations with

$$A(u, p) = -\frac{\partial}{\partial x} p^T B u$$

and

$$u = B^T p.$$

Geometrically, the pair (x, p) is to be thought of as a point in the cotangent bundle T^*X. In this setup each geodesic through x is generated by a choice of $p(0) \in T^*_x X$ but, just as in Riemannian geometry where one does not know *a priori* which values of $\dot{x}(0)$ generate paths over $[0, 1]$ without cut points, here we are not sure *a priori* which values of $p(0)$ generate curves which are free of cut points on $[0, 1]$.

In order to prevent one from attempting to attach intrinsic meaning to an accidental choice of coordinates it is worthwhile to recast these ideas in coordinate free and, while we are at it, global terms. Let X be a manifold and let $\tilde{E}$ be a rank m euclidean vector bundle over X. Let $B: \tilde{E} \to E \subset TX$ be a vector bundle isomorphism. If $\langle \ , \ \rangle$ is the inner product on $\tilde{E}$ then the subbundle of TX defined by E has an inner product which comes from $\langle \ , \ \rangle$. Associated with E is a sequence of derived distributions. Define E^0_x as span $B(x)$ and continue inductively

$$E^{(1)} = \overset{\text{span}}{=} (E^{(0)}_x + [E^{(0)}_x, E^{(0)}_x]), \; E^{(2)} = \overset{\text{span}}{=} (E^{(1)} + [E^{(1)}, E^{(1)}]), \ldots, \text{etc.},$$

where the brackets indicate vector fields which arise as Lie brackets of vector fields in the space indicated. If the dimensions of $E_x^{(i)}$ are, for each i, independent of x then E defines a sequence of derived vector bundles $E^{(0)} \subset E^{(1)} \subset E^{(2)} \subset \cdots$. The condition for the system to be controllable is that this sequence should terminate at TX. Of course E determines, canonically, a dimension m subbundle $E^{\dagger} \subset T^*X$, $E^{\dagger} = \{p \mid p \text{ vanishes on } E\}$. The map $B\colon \tilde{E} \to E$ and the inner product define a map from $T^*X/E^{\dagger}$ into $\tilde{E}$ which is given in coordinates by $p \mapsto B^T p = u$. The pair of equations given above then define a section of the tangent bundle of T^*X. If the controllability condition is satisfied then we get a metric $\rho(\cdot,\cdot)$ on X and we may be sure that any two points in X are joined by a geodesic.

We also point out the following additional result which plays a role later. Suppose that $E^{(1)}$ equals TX. In that case the inner product structure on E can be used to define an inner product on $([E, E] + E)/E$. The idea is analogous to the one whereby an inner product on the space of one forms is used to define an inner product the space of two forms, etc. This goes as follows. Let $b_1, b_2, \ldots, b_m$ be an orthonormal basis for E in some neighborhood $U \subset X$. Any point in $([E, E] + E)/E$ can be then expressed as

$$X = \sum \alpha_{ij}[b_i, b_j] + E.$$

Such a representation is not unique, but among all such representations there is a unique one which minimizes

$$\left(\sum_{ij=1}^{m} \alpha_{ij}^2\right)^{1/2} = \eta(X).$$

This then gives a mapping from $([E, E] + E)/E$ into $\mathbb{R}^{m(m-1)/2}$. It is easily seen to be linear. We define the length of a point in $([E, E] + E)/E$ as the minimum value of $\eta(X)$. It is easy to verify that this defines a norm and that the norm satisfies the parallelogram identity and so it comes from an inner product. Finally, one can check that the norm is independent of the choice of orthonormal basis.

Geodesic Equations

In order to better understand the qualitative behavior of the solutions of the optimal control problem which we introduced in the second section, we now describe a transformation which may be thought of as a partial inverse Legendre transformation. The effect of this transformation is to introduce as many second order equations as possible. Everything here is local.

Given the control equations $\dot{x} = B(x)u$, we then have a subbundle $E = \text{span } B$ in TX. In a neighborhood of any point x_0 we can find an integrable subbundle $\hat{E}$ of TX which is tangent to E at x_0. In local coordinates this amounts to saying that we can write the given equations as

$$\dot{x}_u = B_u u$$

$$\dot{x}_l = B_l u$$

with B_u being an m by m nonsingular matrix and $B_l(x_0) = 0$. For each choice of integrable subbundle $\hat{E}$ tangent to E at x_0 we get such a decomposition of the equations of motion by letting x_u be such that

$$\frac{\partial}{\partial x_u^1}, \ldots, \frac{\partial}{\partial x_u^m}$$

span the integrable subbundle. As noted, E also determines a subbundle $E^\dagger \subset T^*X$, namely the subbundle of one forms which vanish on E. Denoting a typical point in $E^\dagger$ by p_l we can write the equations of the previous section as

$$\dot{x}_u = B_u u$$

$$\dot{x}_l = B_l u$$

$$\dot{p}_u = A_{uu}(u, p_u) + A_{ul}(u, p_l)$$

$$\dot{p}_l = A_{lu}(u, p_u) + A_l(u, p_l).$$

Differentiating the equation $\dot{x}_u = B_u B_u^T p_u + B_u B_l^T p_l$ with respect to time and using the differential equations for p we get a second order equation in x_u. By using $\dot{x}_u = B_u u$ to eliminate u we then end up with a pair of equations of the form

$$\ddot{x}^i + \Gamma^i_{jk}\dot{x}^j\dot{x}^k + \Lambda^i_{jk}\dot{x}^j p^k = 0 \qquad x^i \in \{x^1, x^2, \ldots, x^n\}$$

$$\dot{p}^i + F^i_{jk}\dot{x}^j\dot{x}^k + E^i_{jk}\dot{x}^j p^k = 0 \qquad p^i \in \{p^{m+1}, p^{m+2}, \ldots, p^n\}$$

where the coefficients depend on x but not $\dot{x}$ or p. These equations have to be integrated along with the nonholonomic constraints represented by $\dot{x}_l = B_l B_u^{-1}\dot{x}_u$. The symmetries are as follows: Γ^i_{jk} is symmetric in jk and Λ^i_{jk} is skew symmetric in ij.

Since we did not give a canonical way to choose $\hat{E}$ we cannot attach an intrinsic meaning to any aspect of these equations which is not invariant with respect to that choice. However, given any such choice, B_u defines an inner product on E and hence B_u defines a Riemannian structure on the submanifold passing through x_0 and defined by $\hat{E}$. When we change $\hat{E}$ or x_0 we change this Riemannian structure. We call the original system *reducible* if there exists a choice for $\hat{E}$ such that when we write $\dot{x}_u = B_u u$, B_u is of the form

$$B_u(x_u, x_l) = B_0(x_u)\theta(x_u, x_l)$$

with θ orthogonal. Under this circumstance the Riemannian structure does not vary from leaf to leaf and we can recast the entire problem in the following terms. Given an m-dimensional Riemannian manifold M find the shortest path between two points m_0 and m_1 subject to $n - m$ constraints of the form $y_i(1) = \alpha_i$ where y^i satisfies

$$\dot{y}^i = c_j^i(x, y)u^j.$$

In this case Γ_{jk}^i are just the ordinary Christoffel symbols for this Riemannian manifold.

There exists an entire hierarchy of examples of reducible systems, classified on the basis of the properties of the Riemannian space. For example, the space might be taken to be flat, symmetric, etc. Our prototype problem of section two can be restated as the problem of finding the shortest path between two points $\mathbb{R}^2$ such that the area enclosed by the straight line between the two points and the path has a specified value. This family of special cases is therefore related to the isoperimetric problems in the calculus of variations, and in particular, to the problem of Pappus [8], solved by him more than two thousand years ago.

A Local Canonical Form

Just as the local features of Riemannian geometry are greatly clarified by coordinatizing the manifold by Riemann's normal coordinates, in the present context the local features of the optimal control problem may be revealed by an appropriate choice of coordinates. Specifically, we consider $\dot{x} = B(x)u$ under the replacements

$$x \mapsto x = \Psi(x)$$

$$u \mapsto u = \Theta(x)u,$$

where Ψ is a diffeomorphism and $\Theta(x)$ is an orthogonal matrix. This is the natural group to study because of the role of $\langle u, u\rangle = \langle \Theta u, \Theta u\rangle$ in the optimal control problem. What we will find is that it is possible, under a suitable hypothesis, to get an interesting and useful canonical form with respect to this group of transformations. Everything we do here is local.

To begin with we consider $\dim X = 3$ $\dim \tilde{E} = 2$. What we want to establish is that in this case we can arrange matters so that in a neighborhood of x_0 we have

$$\dot{x}^1 = u^1 + r_1$$

$$\dot{x}^2 = u^2 + r_2$$

$$\dot{x}^3 = u^1x^2 - u^2x^1 + r^3$$

where r_1 and r_2 have vanishing first partials and r^3 has vanishing first partials with respect to x^3 and vanishing first and second partials with respect to x^1 and x^2, all at x_0. Moreover, and this is what justifies the particular choices, any other choice of coordinates which enjoys the same properties is related to the given one by a change of variables whose Jacobian at x_0 has the form

$$\left.\frac{\partial\psi}{\partial x}\right|_0 = \left[\begin{array}{c|c} J_u & 0 \\ \hline 0 & J_l \end{array}\right].$$

Therefore we have an intrinsic definition of a direction $(\partial/\partial x^3)$ which, together with Range B, defines a splitting of the tangent bundle. The prototype problem shows us that this is the direction along which the conjugate points of an associated approximating problem emanate from x_0.

To begin the proof of these assertions consider a system

$$\dot{x}^i = u^i + \gamma^i_{jk} x^j u^k; \qquad i, j, k = 1, 2, \ldots, m.$$

As is well known, any m by m by m array such as γ^i_{jk} can be expressed as a sum $q^i_{jk} + \omega^i_{jk}$ with $q^i_{jk} = q^i_{kj}$ and $\omega^i_{jk} = -\omega^k_{ji}$. By changing coordinates according to

$$x^i \mapsto x^i - \tfrac{1}{2} q^i_{jk} x^j x^k$$

$$u^i \mapsto \theta^i_j u^j,$$

where $\theta = \exp(\Omega(x))$ and $\Omega(x) = (\omega^i_{jk} x^j)$, we arrive at a system

$$\dot{x}^i = u^i + r^i$$

for which all the first partials of r^i vanish at 0. Let's call this a "type one" reduction.

We now consider the x_l equations. For notational reasons we write x and y instead of x_u and x_l. Consider then

$$\dot{x}^i = u^i, \qquad i = 1, 2$$

$$y^i = \alpha^i_{jk} y^j u^k + \beta^i_{jk} x^j u^k + q^i_j u^j$$

where q^i_j have vanishing first partials with respect to x and y at zero.

Split β^i_{jk} up as $\beta^i_{jk} = \bar{\beta}^i_{jk} + \hat{\beta}^i_{jk}$ with the former being symmetric with respect to jk and the latter skew symmetric with respect to the same indices. Notice that if we replace y by $y^i - \tfrac{1}{2}\bar{\beta}^i_{jk} x^j x^k - \alpha^i_{jk} y^j x^k$ then

$$\dot{y}^i = \hat{\beta}^i_{jk} x^j u^k + \hat{q}^i_j u^j$$

where q^i_j have vanishing first partials with respect to x and y at zero. Let's call this a "type two" reduction.

Using a type one reduction followed by a type two reduction we can arrange matters so that the dim $X = 3$, dim $\tilde{E} = 2$ system looks, in a neighborhood of $x = 0$, like

$$\dot{x}^1 = u^1 + r^1$$

$$\dot{x}^2 = u^2 + r^2$$

$$\dot{y} = u^1 x^2 - u^2 x^1 + q^1(x^1, x^2) u^1 + q^2(x^1, x^2) u^2 + r^3$$

where r^3 has the property mentioned above: its first and second partials with respect to x^1 and x^2 vanish at zero, and its first partial with respect to y vanishes at zero. We need to eliminate the quadratic terms q^1 and q^2. To this end substitute

$$y \mapsto y - h(x)$$

with $h(x)$ cubic in x and selected in such a way as to put the expression for $\dot{y}$ in the form

$$\dot{y} = u^1x^2 - x^2u^1 + \alpha x^1yu^1 + \beta x^2yu^2.$$

It is easy to see that such an h exists. After the further substitution

$$y \mapsto y - \alpha x^1 y - \beta x^2 y$$

we have

$$\dot{x}^1 = u^1 + r^1$$

$$\dot{x}^2 = u^2 + r^2$$

$$\dot{y} = u^1(x^2 + \alpha x^1x^2 + \alpha y) - u^2(x^1 - \beta x^1x^2 + \beta y) + r^3.$$

The final reduction to the canonical form is now effected by the substitutions

$$x^1 \mapsto x^1 + \alpha y + \alpha x^1x^2 + 2\beta(x^2)^2$$

$$x^2 \mapsto x^2 + \beta y - \beta x^1x^2 - 2\alpha(x^1)^2$$

$$\begin{bmatrix} u_1 \\ u_2 \end{bmatrix} \mapsto \exp\begin{bmatrix} 0 & \alpha x^2 - \beta x^1 \\ -\alpha x^2 + \beta x^1 & 0 \end{bmatrix}\begin{bmatrix} u^1 \\ u^2 \end{bmatrix}.$$

The statement about the form of the Jacobian at zero may be verified by noticing that the linear transformation which defines the above transformation on (x^1, x^2) has no z component if and only if the system is initially in the desired canonical form.

Based on these techniques we can prove the following theorem.

Theorem 1. *Given* $\dot{x} = B(x)u$ *with* $\dim u = m$ *and* $\dim x = m(m+1)/2$ *and given that* $E^{(1)}$ *spans* $\mathbb{R}^{m(m+1)/2}$ *we can choose coordinates* $(x^1, x^2, \ldots, x^m, y^{1,2}, y^{1,3}, \ldots, y^{m-1,m})$ *in a neighborhood of a given point, say* $x = 0$, *so that the equations take the form*

$$\dot{x}^i = u^i + r^i, \qquad i = 1, 2, \ldots, m$$

$$\dot{y}^{ij} = u^ix^j - u^jx^i + r^{ij}, \qquad i, j = 1, 2, \ldots, m, \quad i < j$$

where the r^i *and* r^{ij} *have vanishing first partials with respect to* x *and* y *and in addition* r^{ij} *has vanishing second partials with respect to* x^i *and* x^j. *Moreover, given any second set of coordinates with this property it follows that the Jacobian of the diffeomorphism which relates them has the block diagonal form*

$$J = \begin{bmatrix} J_{xx} & 0 \\ 0 & J_{yy} \end{bmatrix}$$

when evaluated at zero.

In the next section we will analyze in detail the system defined by this canonical form without the remainder terms. This analysis will explain, in part, the hypothesis that $n = m(m+1)/2$. In fact, even the approximating problem may display a certain lack of robustness with respect to the location of the conjugate points unless this condition on the dimension is satisfied.

Model Spaces

Based on the claim of the previous theorem, the systems of the form

$$\dot{x}^i = u^i, \qquad i = 1, 2, \ldots, m$$

$$\dot{y}^{ij} = u^i x^j - u^j x^i, \qquad i, j = 1, 2, \ldots, m$$

assume a special importance. In this section we explore the associated geodesics. What we will show is that the optimal solution has a remarkably simple structure. One might even think of this class of systems as being the appropriate analog of the flat Riemannian spaces in the present context.

There are many possible notational schemes: we begin with one which is control theoretic and mention at the end an alternative based on differential forms. Consider $x \in \mathbb{R}^m$ and $Y \in o(m)$, the set of n by n skew symmetric matrices. The control system is

$$\dot{x} = u$$

$$\dot{Y} = xu^T - ux^T.$$

It is easy to see that this system is controllable on $\mathbb{R}^n \times o(n)$ and that this is equivalent to the problem defined above. If we are to minimize

$$\eta = \int_0^1 \langle u, u \rangle \, dt$$

subject to fixed boundary conditions $x(0) = 0$, $x(1) = s$, $Y(0) = 0$; if $Y(1) = S$, then an elementary Lagrange multiplier argument shows that there exists a skew symmetric matrix Λ such that

$$\dot{u} + \Lambda u = 0.$$

Thus

$$\dot{x} = u = e^{\Lambda t} a,$$

and the corresponding value of η is just $\|a\|$. Since $x(0) = 0$ and $Y(0) = 0$ we can also write

$$x(t) = e^{\Lambda t} b - b$$

and

$$Y(1) = \int_0^1 ((e^{\Lambda t} b - b) b' \Lambda' e^{\Lambda' t} - e^{\Lambda t} \Lambda b (b' - b' e^{\Lambda' t})) \, dt.$$

The optimal trajectories to points on the set $x(1) = 0$ are especially interesting. In this case the expression for Y simplifies to

$$Y(1) = \int_0^1 e^{\Lambda t}(bb'\Lambda' - \Lambda bb')e^{\Lambda' t}\, dt.$$

Theorem 2. *The control $u(\cdot)$ defined on $[0, 1]$ which minimizes*

$$\eta = \int_0^1 \langle u, u \rangle \, dt$$

for

$$\dot{x} = u; \qquad x(0) = 0 \qquad x(1) = 0$$

$$\dot{Y} = xu^T - ux^T; \qquad Y(0) = 0 \qquad Y(1) = Y$$

satisfies

$$\dot{u} + \Lambda u = 0$$

for some skew-symmetric matrix Λ. The associated value of ρ is given by

$$\rho((0, 0), (0, Y)) = \lambda_1 + 2\lambda_2 + 3\lambda_3 + \cdots + r/2\lambda_r$$

where $\pm i\lambda_1, \pm i\lambda_2, \ldots, \pm i\lambda_r$ are the eigenvalues of Y listed in decreasing order according to the size of the imaginary part. Any two optimal controls u_1 and u_2 transferring $(0, 0)$ to (x, Y) are related by $u_1 = \theta u_2$ for some orthogonal matrix θ such that $\theta Y \theta' = Y$. The point (x, Y) is conjugate to the point $(0, 0)$ if, and only if, x belongs to an invariant subspace of Y which is not in the complement of Ker Y.

Proof. The first observation is that the range space of the operator

$$W = \int_0^1 e^{\Lambda t} bb' e^{\Lambda' t}\, dt$$

is the same as that of $(b, \Lambda b, \ldots, \Lambda^{m-1} b)$ and that the dimension of this range space is upperbounded by the number of distinct eigenvalues of Λ. Second, if $e^{\Lambda} b = b$ then Λ must have at least rank W eigenvalues of the form $2\phi\pi i$ with ϕ an integer since b is necessarily a linear combination of eigenvectors corresponding to such eigenvalues. Finally, if θ is any orthogonal matrix we have $\rho((0, 0), (0, Y)) = \rho((0, 0), (0, \theta' Y \theta))$. We may, therefore, understand the general situation by understanding the case where Λ is of the form

$$\Lambda = \begin{bmatrix} 0 & \omega_1 & & & \\ -\omega_1 & 0 & & & \\ & & 0 & \omega_2 & \\ & & -\omega_2 & 0 & \\ & & & & \ddots \end{bmatrix}$$

with $\omega_k = 2\pi\phi_k$ and no ϕ_k is repeated. In this case a calculation shows that

$$W\Lambda - \Lambda W' = \begin{bmatrix} 0 & (b_1^2 - b_2^2) & 0 & 0 & \cdots \\ (-b_1^2 - b_2^2) & 0 & 0 & 0 & \cdots \\ 0 & 0 & 0 & (b_3^2 - b_4^2) & \cdots \\ 0 & 0 & -(b_3^2 - b_4^2) & 0 & \cdots \\ & & \cdots & \cdots & \end{bmatrix}.$$

But this makes the solution obvious. Since the cost is $\|\Lambda b\|$ we need to pick $\{\phi_1, \phi_2, \ldots, \phi_{m/2}\}$ to be $\{1, 2, \ldots, m/2\}$ if m is even and $\{0, 1, \ldots, (m-1)/2\}$ if m is odd. The total cost will then be expressible in terms of eigenvalues of Y. Say that the eigenvalues of Y with positive imaginary parts are $i\lambda_1, i\lambda_2, \ldots, i\lambda_r$ listed in decreasing size of the imaginary part. Since the eigenvalues of $W\Lambda - \Lambda' W$ are $(b_1^2 - b_2^2)/\phi_1$, etc., we see that the minimum cost is just

$$\eta^* = \lambda_1 + 2\lambda_2 + 3\lambda_3, \ldots, r\lambda_n, \qquad r \leq m/2.$$

As for the lack of uniqueness of u, of course $\dot{x} = u$ implies $\theta\dot{x} = \theta u$ and so θu and u accomplish the same transfer as long as $\theta' Y \theta = Y$. On the other hand, in view of the specific form of the optimal control we see that any two optimal u's which steer $(0, 0)$ to (x, Y) must be so related.

It is worth remarking that while $\rho((0, 0), (0, \alpha Y)) = |\alpha|\, \rho((0, 0), (0, Y))$ it does not define a norm on the space of skew symmetric matrices because the unit ball is not convex. The geometry of the unit ball is partly explained by the remark that the line segments in its boundary correspond to certain line segments in a Cartan subalgebra of $So(n)$.

If we consider a more general version of this problem whereby we wish to control

$$\dot{x} = u$$

$$\dot{y}^i = x^T \Omega_i u, \qquad i = 1, 2, \ldots, r$$

with the Ω_1 skew-symmetric, then it is no longer true that the conjugate points have such a nice structure; in fact, even the connected subset which is conjugate to $(0, 0)$ and contains $(0, 0)$ need not admit the structure of a manifold in any natural way.

There are two possible generalizations of this problem which are interesting and have been investigated in special cases. The first concerns the possibility that it is not $E^{(1)}$ which spans TX but rather some higher $E^{(i)}$. The second concerns isoperimetric problems which are not based on flat spaces but rather spaces of constant curvature.

Let $\Lambda(\mathbb{R}^m)$ denote the 2^m-dimensional Grassmann algebra. Recall that $\Lambda(\mathbb{R}^m)$ splits as the sum of $m + 1$ vector spaces $\Lambda^0(\mathbb{R}^m) + \Lambda^1(\mathbb{R}^m) + \cdots + \Lambda^m(\mathbb{R}^m)$, the pth of which is of dimension $\binom{m}{p}$; $\Lambda^p(\mathbb{R}^m)$ is called the space of p-forms. There is an antisymmetric multiplication in $\Lambda(\mathbb{R}^m)$ denoted by $\wedge$ and called exterior multiplications; it respects the above decomposition in that

$$\wedge : \Lambda^p(\mathbb{R}^m) \times \Lambda^q(\mathbb{R}^m) \to \Lambda^{p+q}(\mathbb{R}^m).$$

Now consider a control system for which $u \in \Lambda^1(\mathbb{R}^n)$, $x \in \Lambda(\mathbb{R}^n)$ and

$$\dot{x}_0 = 0$$

$$\dot{x}_1 = u \wedge x_0$$

$$\cdots\cdots$$

$$\dot{x}_p = u \wedge x_{p-1}.$$

If we set $x_0 = 1$ and delete the first equation, this is a system for which $E^{(i)} = \Lambda^1(\mathbb{R}^m) + \cdots + \Lambda^i(\mathbb{R}^m)$. Above we considered the special case $p = 2$.

A second kind of generalization which yields interesting results concerns systems for which X is principle bundle over a Riemannian space M, u takes on values in TM and the equations of motion are of the form

$$\dot{m} = u$$

$$\dot{Y} = \sum u^i \Omega^i(m) Y$$

where Y is some representation of the group, the $\Omega^i(m)$ belong to the appropriate Lie algebra. The special case of an S^1 bundle over S^2 has been investigated, by J. Baillieul [5] and N. Gunther and T. Goodwille [unpublished]. □

A Second Order Operator

Considerations having to do with stochastic differential equations containing m-dimensional Wiener processes lead, under an appropriate hypothesis, to a naturally defined second order partial differential operator associated with our basic problem. The resulting operator is a generalization of the Laplace–Beltrami operator; it will be hypoelliptic but typically not elliptic.

Recall that an m-dimensional Wiener process w has a $0(m)$ invariance in the sense that the statistical properties of the solution of an Itô equation

$$dx = f(x)\, dt + G(x)\, dw$$

are identical with those of the Itô equations

$$dx = f(x)\, dt + G(x)\theta(x)\, dw$$

where $\theta(x)$ is an orthogonal matrix depending smoothly on x. This $0(m)$ invariance means that the same group of transformations investigated in connection with the local canonical form is relevant here as well.

Given $\dot{x} = B(x)u$ as in the second section we define

$$f^i(x) = -\frac{1}{2}\frac{\partial b^i_j}{\partial x_k} b^k_j \quad \text{(summation convention)}$$

and consider the stochastic equation in the sense of Itô.

$$dx = f(x)\, dt + B(x)\, dw.$$

Together with this equation we consider the equation for the evolution of the associated probability density function $\rho(t, x)$ which is

$$\frac{\partial \rho}{\partial t} = -\frac{\partial}{\partial x^i} f^i(x)\rho + \frac{1}{2}\frac{\partial}{\partial x^i}\frac{\partial}{\partial x^j} b^i_k b^j_k \rho$$

or

$$\frac{\partial \rho}{\partial t} = L_+ \rho.$$

The operator L_+ is not, however, invariantly defined because the probability density ρ is the density with respect to the measure $dx_1, dx_2, \ldots, dx_n$ and, when we change coordinates in X, this transforms by multiplication by the determinant of the Jacobian. The effect on L_+ is therefore

$$L_+ \mapsto \psi^{-1} L_+ \psi$$

where ψ is the determinant of the Jacobian.

If the underlying manifold has a Riemannian structure on it then it has a natural measure, namely $\sqrt{\det G}\, dx_1\, dx_2 \ldots dx_n$ where G is the metric tensor. In that case it may be verified that the operator defined by

$$\frac{1}{(\det G)^{1/2}} L_+ (\det G)^{1/2}$$

is the standard Laplace–Beltrami operator. Thus to get an invariantly defined operator in the present context it is enough to single out a set of diffeomorphisms which are related by transformations whose Jacobians are constant.

Based on the work we have done we are in a position to identify a suitable subset of the set of all diffeomorphisms in the following case. Suppose that for $\dot{x} = B(x)u$ we have $E^{(1)} = TX$ and suppose that $\dim X = m(m+1)/2$ so that Theorem 1 applies. We have a splitting of the tangent space at each point, $T_x X = E_x + F_x$. We also have a method of constructing an inner product on $([E, E] + E)/E$. However, in view of the given decomposition of T_x it can be naturally identified with $E \oplus ([E, E] + E)/E$. Since both these factors have euclidean structures we have obtained from the euclidean structure on E a euclidean structure on $T_x X$. What role this might have in the study of the original problem remains to be investigated, however it does let us define a volume form on $T_x X$ and hence singles out an invariant second order operator.

As remarked at the end of the second section, the second order operator

$$L_+ = \left(\frac{\partial}{\partial x} + y\frac{\partial}{\partial z}\right)^2 + \left(\frac{\partial}{\partial y} - x\frac{\partial}{\partial z}\right)^2$$

plays the role of the heat operator on the metric space defined by $\dot{x} = u$, $\dot{y} = v$, $\dot{z} = xv - yu$. A calculation shows that it is also the second order operator defined by the above procedure. It is hypoelliptic but not elliptic.

In view of the many interesting results which relate the magnitude of the eigenvalues of a Laplace–Beltrami operator to the lengths of geodesics on a compact Riemannian manifold it is natural to expect that this would be a fruitful area of study in the present context. In an earlier paper [9] the spectrum of the Fokker–Planck operator was calculated for a class of problems which fit our framework and the spectrum was related, in a rough way, to the controllability of the systems. The time it takes a Fokker–Planck equation to reach a steady state is of some interest in applications and this is related to the spectrum of the Fokker–Planck equation; perhaps the time is right for a more general study of this type.

References

1. P. Bergmann, *Introduction to the Theory of Relativity*, Prentice–Hall, Englewood Cliffs, NJ, 1942.
2. R. Arnowitt and P. Nath, *Gauge Theories and Modern Field Theory*, M.I.T. Press, Cambridge, MA, 1976.
3. H. Hermes, Attainable sets and generalized geodesic spheres, *J. Diff. Eqs.* **3** (1967), 256–270.
4. R. Hermann, Geodesics of singular Riemannian metrics, *Bull. AMS* **79** (4) (July 1973), 780–782.
5. J. Baillieul, Some optimization problems in geometric control theory, Ph.D. Thesis, Harvard Univ., Cambridge, MA, 1975.
6. W. L. Chow, Über Systeme Von Linearen Partiellen Differentialgleichungen erster Ordnung, *Math. Ann.* **117** (1939), 98–105.
7. S. Kobyayashi, On conjugate and cut Loci, in *Studies in Global Geometry and Analysis*, S. S. Chern, Ed., MAA/Prentice–Hall, Englewood Cliffs, NJ, 1967.
8. C. Carathéodory, *Calculus of Variations and Partial Differential Equations of the First Order*, Holden-Day, San Francisco, CA, 1967.
9. R. W. Brockett, Lie algebras and Lie groups in control theory, in *Geometric Methods in System Theory*, D. Q. Mayne and R. W. Brockett, Eds., Reidel Publishing, Dordrecht, The Netherlands, 1973, pp. 43–82.

On Certain Topological Invariants Arising in System Theory

Christopher I. Byrnes*† and Tyrone E. Duncan**‡

Introduction

This paper is an extension of the lecture presented by the first author at the conference, New Directions in Applied Mathematics, held at The Cleveland Museum of Art and at The Case Western Reserve University on the occasion of that university's centennial anniversary. This author would like to thank the organizers of the conference, Professors Peter Hilton and Gail Young, for their cordial invitation to join in CWRU's celebration as well as for the opportunity to give such a lecture at a time when the mathematics of control systems is expanding quite rapidly on several exciting frontiers. The lecture itself covered, roughly speaking, the material treated in sections 1, 3, and 5 of the present paper and was intended to give an indication of the kind of research which is currently going on in the application of geometry and topology to the problems and theory of linear systems. These topics included a survey of known results and of joint work with the second author, and with R. W. Brockett (this has been reported in more detail elsewhere [10]).

* Research partially supported by the National Aeronautics and Space Administration under Grant NSG-2276, the National Science Foundation under Grant NEG-79-09459, and the Air Force Office of Scientific Research under Grant AFOFR-81-0054.

† Department of Mathematics and Division of Applied Sciences, Harvard University, Cambridge, MA 02138.

** Research partially supported by the U.S. Air Force Office of Scientific Research Grant 77-3177, by the U.S. Office of Naval Research under the Joint Services Electronics Program Contract N00014-75-0648 at Harvard, and as a guest of the SFB 72 of the Deutsche Forschungsgemeinschaft, Bonn.

‡ Department of Mathematics, University of Kansas, Lawrence, KS 66045.

Specifically, these sections deal with two not unrelated aspects of such applications. First, the study of system theoretic invariants of a particular (or perhaps generic) linear system as invariants which arise from a topological, especially homotopy-theoretic, interpretation of the system. This point of view was first systematically studied by Hermann and Martin in an innovative paper ([30]) which interpreted a linear system as a holomorphic map from $\mathbb{CP}^1$ to a suitable Grassmannian, interpreting then several system theoretic invariants as homotopy and holomorphic invariants of rational curves. In this paper and a sequel [17] we consider a refinement of the work of Hermann–Martin in the case where the system possesses a symmetry, perhaps imposed by the physical nature of the system. Explicitly, in this paper we study topological invariants of rational *symmetric* matrix-valued functions—these arise, for example, in the mathematical system theory of linear electric networks. The Hermann–Martin point of view has just recently begun to turn out several results which had not been obtained by the more standard system theoretic techniques. For example, in section 3 we present some positive results on the problem of arbitrarily tuning the natural frequencies of a linear, input–output system, using output feedback. This is a classical open problem in control theory, which turns out in the present setting to boil down to a rather well studied problem in the Schubert calculus—in one language it amounts to calculating the projective volume of a Grassmannian, imbedded by the Plücker imbedding.

The problem of tuning the natural frequencies of a system by output feedback is a problem which is posed, and fits, most naturally in what control theorists would call "the frequency domain." In the past decade there have been several spectacular strides made in basic frequency domain questions (see for example the monograph [48]) and the vista which is now emerging is that such questions are intimately tied up with the geometry of Riemann surfaces, in the simplest nontrivial cases, and of higher dimensional projective varieties in the general cases. In Section 3, we present one of these connections and refer the reader to [10] and [48] for others.

The second area which was touched upon lightly in this lecture is the study of topological properties of spaces (in fact, very often, manifolds) of systems, defined for example by setting certain invariants of the system constant. After a lull which followed the initial investigations on the global analysis of linear systems, research in this area is rapidly turning up exciting results both from the point of view of topology and of applications to system theory. For example, it is known that the homotopy groups of certain of these spaces approximate, in one case, the higher homotopy groups of the 2-sphere and, in another case, the stable homotopy of the unitary groups. In Section 6, we shall give a new result on the existence of "globally convergent" vector fields on a certain class of manifolds of systems, based on G. Segal's calculations concerning the homotopy group of these spaces. This result has potential application to, and in fact was motivated by, problems in the area of system identification.

In addition to extensions of the topics mentioned in the lecture, we have also included (in Section 2) a problem of an arithmetic nature. The problem we pose here is motivated by the problem of realizing, by "state-space equations," the impedance of an RLC network containing transmission lines and turns out, essentially because we are dealing with symmetric systems, to be equivalent to the problem of classifying quadratic forms—or rather quadratic modules—over real polynomial rings. This problem is sometimes referred to as the "quadratic Serre problem," and we make use of results due to Harder, Ojanguren and Parimala to give a partial solution to the problem at hand (see Tables 2.1 and 2.2). A more traditional number theoretic version, over $R = \mathbb{Z}$, was treated in [15]. Aside from this reference, we know of only one other reference ([12]) which deals explicitly with a number theoretic problem in system theory. Although this aspect of linear system theory has not been extensively studied, it is our feeling that there does exist a potentially fruitful number theoretic aspect of algebraic system theory.

The study of several new problems which are characteristic of these three topics of research, together with the applications given in Sections 3 and 6, constitute the bulk of the present manuscript. In order to make the presentation reasonably self-contained, we have also developed some system theoretic preliminaries—these are contained in Sections 1 and 2, and to a certain extent this inclusion dictated the order in which we have presented the material.

We have also included an appendix which gives new, more streamlined proofs of some of the basic results in the global analysis of linear systems. These include, for example, proofs that the set of systems with a fixed number of inputs, outputs and degrees of freedom is naturally a manifold. Since this manifold is describable in at least three different ways, and since there is no published proof that these three descriptions give rise to the same smooth algebraic manifolds, we have also included a rather detailed proof of this fact ("realization with parameters") and have derived statements concerning the elementary topological and geometric properties of these spaces as corollaries.

It is a pleasure to thank Roger Brockett and Bob Hermann for several interesting conversations and suggestions, particularly at the early stages of this work. We would also like to thank Peter Falb and Paul Fuhrmann, who read an earlier version of this manuscript, for the valuable criticisms and suggestions which they shared with us.

1. Topological Invariants of Linear Electrical Circuits

Consider a linear electrical network with positive, constant capacitances and inductances, but which includes a finite number m of current sources. Thus, the network behaves as an input–output device, or a "black box," transforming an applied current $u(t)$ into a resulting voltage $y(t)$, where $u(t_0)$, $y(t_0)$ are real m-vectors. Assuming for the moment that the network has no purely imaginary natural frequencies, one knows ([7], p. 101) from the theory of ordinary differential equations that, for each sinusoidal input current

$$u(t) = u_0 \sin \omega t$$

with frequency ω, there exists a unique periodic response, which may be expressed as:

$$y(t) = G_1(i\omega)u_0 \sin \omega t + G_2(i\omega)u_0 \cos \omega t.$$

By linearity,

$$G(i\omega) = G_1(i\omega) + iG_2(i\omega) \tag{1.1}$$

provides a complete input–output description, in the "frequency domain," of the system. One may also describe such a network by a system of differential equations, and from this description it follows that G is rational in $s = i\omega$, vanishing at ∞, and hence (1.1) may be analytically continued for all but a finite number of frequencies $s \in \mathbb{CP}^1$. If one denotes by $\mathbf{U} = \mathbb{C}^m$, $\mathbf{Y} = \mathbb{C}^m$ the vector spaces of complex inputs and outputs, then following Hermann–Martin ([30]), one may define the holomorphic map

$$\mathbf{G}: \mathbb{CP}^1 \to \mathbf{Grass(m, U \oplus Y)}, \quad \text{via}$$

$$s \mapsto \text{graph } G(s) \subset \mathbf{U} \oplus \mathbf{Y}. \tag{1.2}$$

Of course, the definition in (1.2) must be extended across the poles of $G(s)$, which are inessential singularities. Indeed, the number of poles of $G(s)$, counted with multiplicity, is equal to the degree of $\mathbf{G}$

$$[\mathbf{G}] \in \pi_2(\mathbf{Grass(m, U \oplus Y)}) \simeq \mathbf{Z}. \tag{1.3}$$

For, $\deg_{\mathbb{C}}(\mathbf{G})$ may be computed as the intersection number

$$(\mathbf{G}(\mathbb{CP}^1): \boldsymbol{\sigma}(\mathbf{Y})) = \deg_{\mathbb{C}}(\mathbf{G}) \tag{1.4}$$

where

$$\boldsymbol{\sigma}(\mathbf{Y}) = \{W: \dim(W \cap \mathbf{Y}) \geq 1\} \subset \mathbf{Grass(m, U \oplus Y)} \tag{1.5}$$

is the Schubert hypersurface associated to Y. Now, to say graph $G(s_0) \subset \mathbf{U} \oplus \mathbf{Y}$ is the graph of a linear function mapping $\mathbf{U}$ to $\mathbf{Y}$ is to say graph $G(s_0)$ is transverse to Y. Therefore, the nontrivial contributions to (1.4) arise from the poles of $G(s)$ counted with multiplicity. The basic identity (1.5) is the starting point for a rather striking application, discussed in Section 3, of the Schubert calculus to a classical problem in feedback control. Returning to the present theme, this intersection number is exactly the minimum number of capacitors and inductors occurring in any network with frequency response $G(s)$; that is,

$$\deg_{\mathbb{C}}(\mathbf{G}) = \# \mathbf{C} + \# \mathbf{L}. \tag{1.6}$$

This identity is well known, following from the Hermann–Martin calculation of $\deg_{\mathbb{C}}(\mathbf{G})$ and from a beautiful tool—known as realization theory, adapted to the symmetric case at hand—which relates frequency domain descriptions of a system to state-space representations.

More explicitly, under quite general hypotheses, one may describe such a network by differential equations involving the rates of change in the voltage across the capacitors and the current in the inductors. Indeed, if one disconnects the current sources and allows nonlinear resistors, then according to Brayton–Moser ([6]) the equations

$$\begin{aligned} L_\rho \frac{di_\rho}{dt} &= \frac{\partial P}{\partial i_\rho} \qquad (\rho = 1, \ldots, \#\, \mathbf{L}) \\ C_\sigma \frac{dv_\sigma}{dt} &= -\frac{\partial P}{\partial v_\sigma} \qquad (\sigma = 1, \ldots, \#\, \mathbf{C}) \end{aligned} \tag{1.7}$$

describe the dynamics of such a circuit, provided the resistive elements may be extracted from the active elements. Thus the equations (1.7) may be regarded as a "gradient" flow, with respect to the indefinite metric

$$-\sum_{\rho=1}^{\# L} L_\rho (di_\rho)^2 + \sum_{\sigma=1}^{\# C} C_\sigma (dv_\sigma)^2 \tag{1.7$'$}$$

on the state manifold M, with coordinates (i_ρ, v_σ). It is noted in [6] that (1.7) differs from the Hamiltonian systems in particle dynamics in that the mixed potential P contains dissipative terms and in that the equations (1.7) preserve their form under any change of coordinates which leaves the indefinite metric (1.7)′ invariant.

Of course, in the linear case the Brayton–Moser equations assume the familiar form

$$\frac{dx}{dt} = Ax, \qquad AI_{p,q} = I_{p,q}A^t \tag{1.8}$$

where

$$I_{p,q} = \operatorname{diag}(\varepsilon_i) \quad \text{with}$$

$$\varepsilon_i = \begin{cases} +1 & i = 1, \ldots, p \\ -1 & i = p+1, \ldots, p+q. \end{cases}$$

Now, under the hypothesis that the resistive elements may be extracted from the active elements, Youla–Tissi ([57]) have shown that the circuit dynamics for a linear network with current sources may be described by equations of the form

$$\begin{aligned} \frac{dx}{dt} &= Ax + Bu \\ y &= Cx \end{aligned} \tag{1.9}$$

where the internal symmetry properties

$$\begin{aligned} AI_{p,q} &= I_{p,q}A^t \\ I_{p,q}B &= C^t \end{aligned} \tag{1.10}$$

are satisfied. Thus, (1.9) and (1.10) give a "state-space," or internal, realization of such a circuit. Moreover, one may recover the frequency domain data as the Laplace transform of the fundamental solution to (1.9),

$$\hat{y}(s) = G(s)\hat{u}(s) = C(sI - A)^{-1}B\hat{u}(s). \tag{1.11}$$

Now, a basic result due to Youla and Tissi ([57]), which is useful in the synthesis of networks as well as in applications of numerical analysis, asserts that such "internally symmetric" realizations exist for any real symmetric matrix-valued rational function $G(s)$, vanishing at ∞. Equation (1.11), which is an analogue of Ohm's Law, yields a proof of (1.6) as well as a proof that the frequency response function $G(s)$ is a real, rational, matrix-valued function, vanishing at ∞. One may also deduce by counting poles that $\deg_{\mathbb{C}}(\mathbf{G})$ is a lower bound for the dimension of any realization (1.9) of $G(s)$, satisfying (1.10) or not.

Definition 1.1. A triple (A, B, C) satisfying (1.10), with (1.9) understood, will be referred to as an internally symmetric realization of $G(s)$. If $G(s)$ is $m \times m$ and $\deg_{\mathbb{C}}(\mathbf{G}) = n$, A is $n \times n$ while B and C^t are $n \times m$, then (A, B, C) will be called a minimal, internally symmetric realization and the subset of $\mathbb{R}^{\mathbf{n^2+2nm}}$ consisting of such triples will be denoted by $\widetilde{\mathbf{Rat}}(\mathbf{p}, \mathbf{q}; \mathbf{m})$. The union, over all signature matrices $I_{p,q}$, of these spaces will be denoted by

$$\widetilde{\mathbf{Rat}}(\mathbf{n}; \mathbf{m}) = \bigcup_{\mathbf{p}+\mathbf{q}=\mathbf{n}} \widetilde{\mathbf{Rat}}(\mathbf{p}, \mathbf{q}; \mathbf{m}).$$

It is an open question, partially solved in Section 2, to give criteria for the existence of such realizations for systems containing time delays, or for systems depending upon parameters. Formulated precisely, the first extension is equivalent to the "quadratic Serre Problem," and the second extension is equivalent to the construction and analysis of the moduli space **Rat(n; m)** of real symmetric $m \times m$ matrix valued rational functions satisfying

$$\deg_{\mathbb{C}}(\mathbf{G}) = n \tag{1.12a}$$

$$\mathbf{G}(\infty) = \mathbf{0}. \tag{1.12b}$$

By our constructions in Section 4, **Rat(n; m)** is a smooth real algebraic manifold of dimension $2nm$, which is a classifying space for systems containing parameters.

When $m > 1$, to say that $G(s)$ is symmetric is to say that the network is reciprocal: the effect of current u_i in the ith current source on the voltage y_j

across the jth current source is the same as the effect of u_j on y_i, reflecting a very basic duality between current and voltage (see [40], Part I, Chap. IX). Indeed, regarding **Y** as **U***, we have for $\mathbf{G} \in \mathbf{Rat(n; m)}$

$$\langle u, G(s_0)v\rangle - \langle G(s_0)u, v\rangle = 0, \tag{1.13}$$

that is, graph $G(s_0)$ is a Lagrangian plane for the standard symplectic form on $\mathbf{U} \oplus \mathbf{U}^*$. Since G is real, the frequency response function gives rise to a map,

$$\mathbf{G}\colon \mathbb{RP}^1 \to \mathbf{LG}(\mathbf{m}, \mathbf{U} \oplus \mathbf{U}^*), \tag{1.14}$$

to the Lagrangian Grassmannian and hence to a *second invariant*

$$\mathrm{Ind}(\mathbf{G}) \in \pi_1(\mathbf{LG}(\mathbf{m}, \mathbf{U} \oplus \mathbf{U}^*)) \simeq \mathbf{Z} \tag{1.15}$$

—the Arnol'd–Maslov index of G. This may be computed ([1], Theorem 1.5) by intersecting with the Maslov cycle $\boldsymbol{\sigma}(\mathbf{U}^*)$

$$\mathrm{Ind}(\mathbf{G}) = (\mathbf{G}(\mathbb{RP}^1)\colon \boldsymbol{\sigma}(\mathbf{U}^*)), \tag{1.16}$$

where $\boldsymbol{\sigma}(\mathbf{U}^*) \subset \mathbf{LG}(\mathbf{m}, \mathbf{U} \oplus \mathbf{U}^*)$ is defined as in (1.5). In particular,

$$\begin{gathered} |\mathrm{Ind}(\mathbf{G})| \leq \deg_{\mathbb{C}}(\mathbf{G}), \quad \text{and} \\ \mathrm{Ind}(\mathbf{G}) \equiv \deg_{\mathbb{C}}(\mathbf{G}) \bmod(2). \end{gathered} \tag{1.17}$$

In accordance with (1.17), we may define the $n + 1$ subspaces of $\mathbf{Rat(n; m)}$:

$$\mathbf{Rat(p, q; m)} = \{\mathbf{G}\colon \mathrm{Ind}(\mathbf{G}) = p - q, \qquad \deg_{\mathbb{C}}(\mathbf{G}) = p + q\}.$$

We shall prove that each $\mathbf{Rat(p, q; m)}$ is a path component of $\mathbf{Rat(n; m)}$

$$\mathbf{Rat(n; m)} = \bigcup_{\mathbf{p}+\mathbf{q}=\mathbf{n}} \mathbf{Rat(p, q; m)}. \tag{1.18}$$

Furthermore, there is a natural C^∞ quadratic form $\mathscr{H}_{\mathbf{G}}$ defined for $\mathbf{G} \in \mathbf{Rat(n; m)}$, such that

$$\mathrm{Ind}(\mathbf{G}) = \mathrm{sgn}(\mathscr{H}_{\mathbf{G}}) \tag{1.19}$$

and such that

$$\mathrm{sgn}(\mathscr{H}_{\mathbf{G}}) = \#\,\mathbf{C} - \#\,\mathbf{L} \tag{1.20}$$

whenever $G(s)$ is the frequency response of an *RLC* circuit, giving another interpretation of (1.17).

If $m = 1$, (1.19) is the Hermite–Hurwitz Theorem ([33]) §3) which was discovered in connection with the Routh–Hurwitz conditions for stability of constant coefficient linear ordinary differential equations. A global proof of the Hermite–Hurwitz Theorem was given by R. W. Brockett in [8] as a corollary to (1.18), observing that each side of (1.19) is a continuous integer-valued function on $\mathbf{Rat}(\mathbf{n}; \mathbb{R}) \equiv \mathbf{Rat(n; 1)}$, and finally by evaluating the identity once on each component.

2. Realization Theory for Delay Differential Systems and the Quadratic Serre Problem

Recall that if $G(s)$ is a $p \times m$ rational matrix-valued function defined over a field k vanishing at ∞, then according to the State Space Realization Theorem ([34], Theorem 3), $G(s)$ admits a factorization (or realization)

$$G(s) = C(sI - A)^{-1}B = \sum_{i=1}^{\infty} \frac{CA^{i-1}B}{s^i} \tag{2.1}$$

defined over k. In particular, G admits a minimal dimension realization in the sense that $A \in \mathcal{M}_n(k)$ with n minimal, and it is clear from (1.4) that

$$n = \deg_{\bar{k}}(\mathbf{G}) \tag{2.2}$$

is this dimension. On the other hand, it is clear from (2.1) that one has

$$\deg_{\bar{k}}(\mathbf{G}) = \operatorname{rank}_k(\mathcal{H}_{\mathbf{G}}) \tag{2.3}$$

where the (i, j)th $p \times m$ block of the Hankel matrix $\mathcal{H}_G$ is defined as

$$(\mathcal{H}_G)_{i,j} = CA^{i+j-2}B. \tag{2.4a}$$

For $m = p = 1$, this is the standard result relating, via the Cayley–Hamilton Theorem, rationality of a generating function to the existence of a recurrence relation on the Taylor, or Laurent, coefficients ([39]).

In continuous-time, the factorization (2.4a) has the interpretation that $G(s)$ is the Laplace transform of the fundamental solution of the system

$$\frac{dx}{dt} = Ax(t) + Bu(t), \qquad x(0) = 0$$

$$y(t) = Cx(t). \tag{2.4b}$$

Definition 2.1. A triple of k-linear maps (A, B, C) which define a minimal realization (2.1) of some $G(s)$ will be referred to as a minimal system of degree $n = \deg_{\bar{k}}(\mathbf{G})$. The subset of $\mathbb{A}^{\mathbf{n^2+nm+np}}$ of minimal systems of degree n will be denoted by $\tilde{\Sigma}^n_{m,p}(k)$.

In fact, the rank condition (2.3) shows that $\tilde{\Sigma}^n_{m,p}(k)$ is Zariski open in $\mathbb{A}^{\mathbf{n^2+nm+np}}$. Minimality is actually quite a bit stronger and from this property one may deduce the second part of the State Space Realization Theorem. That is, a change of basis defined by $T \in \mathbf{GL(n, k)}$ induces a transformation on minimal realizations

$$\alpha(T, (A, B, C)) = (TAT^{-1}, TB, CT^{-1}) \tag{2.5}$$

which leaves the external behaviour of the system, viz. $G(s)$, invariant. Furthermore, using classical invariant theory, one may show in the scalar input–output case that the entries of the Hankel parameters, CA^iB, generate the ring of invariants for the action (2.5)

$$\alpha: \mathbf{GL(n, k)} \times \mathbb{A}^{\mathbf{n^2+nm+np}} \to \mathbb{A}^{\mathbf{n^2+nm+np}}.$$

Now, α restricts to an action

$$\alpha: \mathbf{GL(n, k)} \times \tilde{\boldsymbol{\Sigma}}^{\mathbf{n}}_{\mathbf{m, p}}(\mathbf{k}) \to \tilde{\boldsymbol{\Sigma}}^{\mathbf{n}}_{\mathbf{m, p}}(\mathbf{k}) \tag{2.6}$$

and the State-Space Isomorphism Theorem ([34], Theorem 3) asserts that $(A_1, B_1, C_1), (A_2, B_2, C_2) \in \tilde{\Sigma}^n_{m, p}(k)$ realize the same transfer function if, and only if,

$$(TA_1T^{-1}, TB_1, C_1T^{-1}) = (A_2, B_2, C_2) \tag{2.7}$$

holds for a unique $T \in \mathbf{GL(n, k)}$. In other words,

Proposition 2.2. *The action α defined by (2.6) is free, with orbit space in canonical bijection with $\Sigma^n_{m, p}(k)$—the set of $p \times m$ rational, matrix-valued functions defined over k, satisfying*

$$\mathbf{G}(\infty) = 0 \tag{2.8}$$

$$\deg_{\bar{k}}(\mathbf{G}) = n. \tag{2.9}$$

Finally, minimality is equivalent (over a scalar field k) to reachability and observability of the discrete-time system

$$\begin{aligned} x(t+1) &= Ax(t) + Bu(t) \\ y(t) &= Cx(t) \end{aligned} \tag{2.10}$$

or of the associated continuous-time differential system if k is complete with respect to an absolute value. Reachability is the property that each state x_1 may be reached from the origin along a trajectory of (2.10) by proper choice of input $u(t)$, i.e.,

$$\operatorname{rank}_k(B, AB, \ldots, A^{n-1}B) = n. \tag{2.11}$$

Dually, observability is the property that each state x may be distinguished from the origin by sufficiently long observation, i.e.,

$$\ker(C, CA, \ldots, CA^{n-1}) = (0). \tag{2.12}$$

These are highly desirable properties in any model of a dynamical system, and serve to put minimality in a proper perspective. The success of linear control theory has led to many attempts to extend the theory to more general cases. In many physical applications, electrical networks are distributed, in effect containing significant delays in transmission between certain nodes, and it is reasonable to model such networks by a system

$$\begin{aligned} \frac{dx}{dt} &= A * x(t) + B * u(t) \\ y(t) &= C * x(t) \end{aligned} \tag{2.13}$$

where A, B, C are finite matrices of bounded measures with compact support in $[0, \infty)$ and where the initial data x_0, u_0 are elements of $L^1[-\tau, 0]^{(n)}$ and $L^1[-\tau, 0]^{(m)}$, respectively, with τ the least upper bound of the supports of these measures. In particular, (2.13) exhibits such a system as a system defined over a subring R of the integral domain of Schwartz distributions, and therefore as a system defined over the fraction field K of R. Indeed, early attempts ([56]) at a realization theory for such systems appealed to the realization theory over K. This technique leads of course to quite serious integrality questions (since the inverse of a 1 second delay is a 1 second predictor), which were addressed in a series of papers ([35–36]) by E. Kamen using the extension of realization theory to integrally closed Noetherian domains ([49]).

A problem we address in this section is the existence of canonical (i.e., reachable and observable) internally symmetric realizations for symmetric rational $G(s)$ defined over a polynomial ring $k[Z_1, \ldots, Z_n]$. This is particularly interesting in the case $k = \mathbb{R}$, which arises when the bounded measures in (2.13) are linear combinations of point masses. It may be shown ([14], Proposition 1.1) that in this case (2.13) is definable over a ring $R = k[Z_1, \ldots, Z_n]$, where

$$Z_i * x(t) = x(t - \theta_i), \qquad \theta_i > 0$$

with θ_i rationally independent. Now, in this case, the Laplace transform of the fundamental solution to (2.13) is rational in s, and polynomial in the variables $Z_i = e^{-\theta_i s}$ and the formal questions of realizability over R correspond precisely to the questions of realizability of frequency response data by delay-differential systems. As one might suspect, the existence of internally symmetric realizations defined over R is intimately tied up with the classification of quadratic forms over R.

For simplicity, we shall first consider the scalar case, so that $G(s)$ is a rational function vanishing at ∞ and defined over $R = \mathbb{R}[Z_1, \ldots, Z_n]$, with Laurent expansion

$$G(s) = \sum \frac{h_i}{s^i}, \qquad h_i \in R$$

leading to the (truncated) Hankel matrix

$$\mathscr{H}'_G = (h_{i+j-1})_{i,j=1}^n. \tag{2.14}$$

In (2.14), n is the rank of $\mathscr{H}_G$ over the field $K = \mathbb{R}(Z_1, \ldots, Z_N)$.

Theorem 2.3. *A scalar transfer function $G(s)$ defined over $\mathbb{R}[Z]$ admits a canonical internally symmetric realization if, and only if, $\mathscr{H}'_G$ is nondegenerate.*

PROOF. Suppose, as we may for any scalar $G(s)$, that (2.13) is a reachable and observable realization over R of $G(s)$; i.e.,

$$\operatorname{span}(B, AB, \ldots, A^{n-1}B) = R^{(n)} \tag{2.15}$$

$$\ker(C, CA, \ldots, CA^{n-1}) = (0). \tag{2.16}$$

If (A, B, C) is internally symmetric, i.e., satisfies (1.10), then by (2.4)

$$(B, AB, \ldots, A^{n-1}B)^t I_{p,q}(B, AB, \ldots, A^{n-1}B) = \mathscr{H}'_G \tag{2.17}$$

which proves the necessity of the condition on $\mathscr{H}'_G$. Conversely, a theorem of G. Harder ([37]) asserts that if $\mathscr{Q}$ is a nondegenerate symmetric form defined over $\mathbb{R}[Z]$, then $\mathscr{Q}$ is extended from $\mathbb{R}$ or, equivalently, $\mathscr{Q}$ is conjugate to some $I_{p,q}$. Therefore, the theorem follows from the following result.

Lemma 2.4. *A scalar $G(s)$ admits a canonical internally symmetric realization over $\mathbb{R}[Z_1, \ldots, Z_N]$ if, and only if, the Hankel form $\mathscr{H}'_G$ is extended from $\mathbb{R}$.*

Proof. Of course, we need only prove sufficiency. To say $\mathscr{H}'_G$ is nondegenerate is to say

$$\operatorname{rank} \mathscr{H}_G(M) = n(M) \equiv n, \quad \text{for all } M \in \operatorname{Max} R. \tag{2.18}$$

Now, over any Noetherian integral domain R, it is known ([51], Theorem 2.1) that the condition (2.18) implies the existence of a finitely-generated projective (state) module Q and R-module maps

$$A: Q \to Q, \qquad B: R^{(m)} \to Q, \qquad C: Q \to R^{(p)}$$

satisfying the identity (2.1)

$$G(s) = \sum_{i=1}^{\infty} \frac{CA^{i-1}B}{s^i},$$

and such that the conditions

$$\mathscr{C} = (B, AB, \ldots): \bigoplus^{\infty} R^{(m)} \to Q \quad \text{is surjective} \tag{2.15$'$}$$

$$\mathscr{O} = (C, CA, \ldots)^t: \operatorname{Hom}_R\left(\bigoplus^{\infty} R^{(p)}, R\right) \to \operatorname{Hom}_R(Q, R) \quad \text{is surjective} \tag{2.16$'$}$$

hold. If $R = \mathbb{R}[Z_1, \ldots, Z_N]$, then $Q \simeq R^{(n)}$ by the Quillen–Suslin Theorem and $G(s)$ may be realized by matrices (A, B, C) over R. Moreover, if $G(s)$ is symmetric, then both (A, B, C) and (A^t, C^t, B^t) are realizations of $G(s)$, satisfying (2.15) and (2.16). By the state-space isomorphism theorem, these realizations are related by a unique $T \in \mathbf{GL(n, R)}$ via

$$\begin{aligned} TA &= A^tT \\ TB &= C^t \\ C &= B^tT. \end{aligned} \tag{2.19}$$

By uniqueness, $T = T^t$, and it is straightforward to check the identity

$$(B, AB, \ldots, A^{n-1}B)^t T(B, AB, \ldots, A^{n-1}B) = \mathscr{H}'_G. \tag{2.17$'$}$$

In particular, to say that T is extended from $\mathbb{R}$ is to say $G(s)$ admits an internally symmetric realization by a triple of matrices. Q.E.D.

From Lemma 2.4 we may derive another result for scalar $G(s)$; viz., it is a theorem of M. Ojanguren that over $\mathbb{R}[Z_1, \ldots, Z_N]$ every indefinite nondegenerate form $\mathcal{Q}$ is extended from $\mathbb{R}$, see [46].

Corollary 2.5. *If $G(s)$ is a scalar transfer function defined over $\mathbb{R}[Z_1, \ldots, Z_N]$ with a nondegenerate, indefinite Hankel form $\mathcal{H}'_G$, then $G(s)$ admits a canonical, internally symmetric realization.*

Remark. At present we do not know whether Corollary 2.5 holds in the positive or negative definite cases. S. Parimala has given an example [47] of a 4×4 positive definite form $\mathcal{Q}$ defined over $\mathbb{R}[Z_1, Z_2]$ but whether or not $\mathcal{Q}$ is conjugate to a Hankel form remains open.

In the multivariable setting, of course, the Hankel form $\mathcal{H}'_G$ is a block matrix of rank n and is no longer nondegenerate. In particular, even if $G(s)$ admits an internally symmetric realization the conjugacy (2.17) must be relaxed to the condition (2.18), and the appropriate form of Lemma 2.4 becomes the assertion

"$G(s)$ admits a canonical, internally symmetric realization if, and only if, the symmetric form T in (2.19) is extended from $\mathbb{R}$,"

while one still obtains (2.17)′ and, in particular, the identity

$$\operatorname{sgn} \mathcal{H}'_G = \operatorname{sgn} T.$$

In another direction, this assertion also implies that the question of existence of internally symmetric realizations is, in fact, equivalent to the "quadratic Serre problem."

Proposition 2.6. *Every symmetric transfer function $G(s)$, defined over $R = \mathbb{R}[Z_1, \ldots, Z_N]$ admits a (reachable and observable) internally symmetric realization if, and only if, each nondegenerate symmetric form $\mathcal{Q}$, defined over R, is extended from $\mathbb{R}$.*

PROOF. We have already proved sufficiency, now suppose $\mathcal{Q}$ is given. One may form the transfer function

$$G(s) = (s - I)^{-1}$$

which admits the reachable and observable realization $(I, \mathcal{Q}, I)$. In light of (2.19) with $T = \mathcal{Q}$, $G(s)$ admits such a realization if, and only if, $\mathcal{Q}$ is extended from $\mathbb{R}$. Q.E.D.

Thus we can summarize these results in the following tables, where we have answered the existence questions concerning internally symmetric transfer functions.

Table 2.1 Scalar $G(s)$

	$\mathbb{R}[Z]$	$\mathbb{R}[Z_1, \ldots, Z_N]$, $N \geq 2$
$\mathscr{H}'_G$ pos. or neg. definite	Yes	?
$\mathscr{H}'_G$ indefinite	Yes	Yes

Table 2.2 Multivariable $G(s)$

	$\mathbb{R}[Z]$	$\mathbb{R}[Z_1, \ldots, Z_N]$, $N \geq 2$
$\mathscr{H}'_G$ pos. or neg. semidefinite	Yes	No
$\mathscr{H}'_G$ indefinite	Yes	Yes

3. An Application of the Schubert Calculus to Output Feedback Control

Consider a $p \times m$ frequency response function $G(s)$ having $\deg_{\mathbb{C}}(\mathbf{G})$, n. That is,

$$\hat{y}(s) = G(s)\hat{u}(s), \qquad y \in \mathbb{C}^{\mathbf{p}}, \qquad u \in \mathbb{C}^{\mathbf{m}} \tag{3.1}$$

is the input–output description of a linear dynamical system of order n. If one "feeds back" (a linear function $-Ky$ of) the output as an input, so that u is replaced by $u - Ky$, then the frequency response of the new "closed-loop" feedback system is given by

$$\hat{y}(s) = G^K(s)\hat{u}(s), \qquad G^K(s) = G(s)(I - KG(s))^{-1}. \tag{3.2}$$

In quite vague terms, the principal question which arises in this context is: what can be done (i.e., which behavioural characteristics can be achieved) using output feedback? In this genre of problems, the question of which (divisors of) poles can be achieved by suitable choice of K is both classical and important. For example, (2.4b), to place the poles in the left-half plane is to stabilize the original system externally by feedback. If $(\lambda_1, \ldots, \lambda_n)$ is the unordered n-tuple of complex poles of $G^K(s)$, denote by $\chi_G(K) \in \mathbb{C}^n$ the vector of coefficients of the (closed-loop characteristic) polynomial

$$p_K(t) = \prod_{i=1}^{n} (t - \lambda_i).$$

In the situation of interest, G and K are taken to be real so that $\chi_G(K) \in \mathbb{R}^{\mathbf{n}}$. Thus, the (real) pole-placement problem is whether

$$\chi_G \colon \mathbb{R}^{\mathbf{mp}} \to \mathbb{R}^{\mathbf{n}}$$

is surjective. It is not hard to check that, if $mp \geq n$, χ_G is a submersion at some $K \in \mathbb{R}^{pm}$ for generic G and therefore, over $\mathbb{C}$, *image* (χ_G) contains nonempty Zariski open subset of $\mathbb{C}^n$, by the dominant morphism theorem (see [29]). Over $\mathbb{R}$, however, it has been shown ([54]) by brute force, that if $m = p = 2$, $n = 4$, *image* χ_G is not dense for generic G.

Consider first the direct problem, to compute $\chi_G(K)$, where we allow complex linear maps K and hence complex values for $\chi_G(K)$. In the setting of Section 1, $\chi_G(0)$ is given by the formula

$$\mathbf{G}^{-1}(\mathbf{G}(\mathbb{C}\mathbb{P}^1) \cap \boldsymbol{\sigma}(\mathbf{Y})) = \chi_G(0) \tag{3.3}$$

where

$$\mathbf{G}: \mathbb{C}\mathbb{P}^1 \to \mathbf{Grass(m, U \oplus Y)}$$

is the rational curve constructed in (1.2) and $\boldsymbol{\sigma}(\mathbf{Y})$ is the Schubert hypersurface defined in (1.5). Thus, if $\mathbf{0}: \mathbf{Y} \to \mathbf{U}$ is the zero linear transformation, then $\mathbf{graph(0)} \subset \mathbf{U \oplus Y}$ is a p-plane, viz. $\mathbf{Y}$, and one may rewrite (3.3) in a more suggestive form

$$\mathbf{G}^{-1}(\mathbf{G}(\mathbb{C}\mathbb{P}^1) \cap \boldsymbol{\sigma}(\mathbf{graph(0)})) = \{\textit{Poles of } G^0(s)\}. \tag{3.3$'$}$$

Indeed, a straightforward calculation shows that, if for a p-plane $\mathbf{V}$ we define $\boldsymbol{\sigma}(\mathbf{V}) \subset \mathbf{Grass(m, U \oplus Y)}$ via

$$\boldsymbol{\sigma}(\mathbf{V}) = \{W: \dim(W \cap V) \geq 1\}, \tag{3.4}$$

then

$$\mathbf{G}^{-1}(\mathbf{G}(\mathbb{C}\mathbb{P}^1) \cap \boldsymbol{\sigma}(\mathbf{graph(K)})) = \{\textit{Poles of } G^K(s)\}. \tag{3.5}$$

Thus, the direct problem is equivalent to the classical construction: *given a rational curve* $\mathbf{C} \subset \mathbf{Grass(p, m + p)}$ *and a (Schubert) hypersurface* $\boldsymbol{\sigma}(\mathbf{V})$ *construct the divisor* $\mathscr{D}$ *on* $\mathbf{C}$ *defined as the geometric intersection* $\mathbf{C} \cap \boldsymbol{\sigma}(\mathbf{V})$.

The inverse problem is also quite classical, viz. given a divisor $\mathscr{D}$ on C to find a hypersurface $\mathbf{H}$ in a given family of hypersurfaces $\mathscr{H}$ such that

$$\mathscr{D} = \mathbf{C} \cap \mathbf{H},$$

but here there is also another rub: C and $\mathscr{D}$ are taken to be defined by real equations, and $\mathbf{H}$ is required to be real as well. For the problem at hand, the family of hypersurfaces is parameterized by the Grassmannian, $\mathbf{Grass(p, U \oplus Y)}$, where the p-plane $\mathbf{V}$ gives rise to the Schubert hypersurface $\boldsymbol{\sigma}(\mathbf{V})$ as defined in (3.4) and where (3.4) may be taken as the definition of the rational function χ_G on the open cell $\mathbf{Grass(p, U \oplus Y)} - \boldsymbol{\sigma}(\mathbf{U}) \cong \mathbb{C}^{mp}$. In this setting the problem of pole-placement is to analyze the (real) behavior of the rational map

$$\chi_G: \mathbf{Grass(p, U \oplus Y)} \to (\mathbb{C}\mathbb{P}^1)^{(n)} \simeq \mathbb{C}\mathbb{P}^n. \tag{3.6}$$

Recall that the space $(\mathbb{C}\mathbb{P}^1)^{(n)}$ of unordered n-tuples of points on $\mathbb{C}\mathbb{P}^1$ is canonically isomorphic to $\mathbb{C}\mathbb{P}^n$. Here, if $\mathbb{C}\mathbb{P}^{n-1}_\infty$ denotes the variety of n-tuples with at least one entry ∞, $\mathbb{C}\mathbb{P}^n - \mathbb{C}\mathbb{P}^{n-1}_\infty \simeq \mathbb{C}^n \simeq (\mathbb{C}^1)^{(n)}$.

Lemma 3.1 ([10]). *If $mp \le n$, for the generic G, χ_G in* (3.6) *extends to an everywhere regular map on the Grassmannian which satisfies $\chi(\sigma(\mathbf{U})) \subset \mathbb{CP}^{\mathbf{n}-1}_{\infty}$. In particular, the maps*

$$\chi_G\colon \mathbf{Grass(p, U \oplus Y)} - \sigma(\mathbf{U}) \simeq \mathbb{C}^{\mathbf{mp}} \to \mathbb{CP}^{\mathbf{n}} - \mathbb{CP}^{\mathbf{n}-1}_{\infty} \simeq \mathbb{C}^{\mathbf{n}} \tag{3.7a}$$

$$\chi_G\colon \mathbb{R}^{\mathbf{mp}} \to \mathbb{R}^{\mathbf{n}} \tag{3.7b}$$

are proper, i.e., for $K \subset \mathbb{C}^{\mathbf{n}}$ compact, $\chi_G^{-1}(K)$ is compact.

Remark. By generic, we mean that the space $\Sigma^{\mathbf{n}}_{\mathbf{m,p}}(\mathbb{C})$ of rational $p \times m$ matrix-values G, vanishing at ∞ and having degree n, is naturally a smooth (irreducible) variety and that the space $\mathbf{V}^{\mathbf{n}}_{\mathbf{m,p}}$ of G's for which χ_G extends is Zariski open. Indeed, in the appendix we shall present a new, rather elementary, proof of the first assertion which also shows that the underlying Euclidean topology on $\Sigma^{\mathbf{n}}_{\mathbf{m,p}}(\mathbb{C})$ is the subspace topology induced by the (Hermann–Martin) inclusion

$$\Sigma^{\mathbf{n}}_{\mathbf{m,p}}(\mathbb{C}) \subset \Omega^{2}_{(\mathbf{n})}(\mathbf{Grass(m, U \oplus Y)})$$

in the 2nd loop space of $\mathbf{Grass(m, U \oplus Y)}$. In particular, $\mathbf{V}^{\mathbf{n}}_{\mathbf{m,p}}$ is connected in the compact-open topology.

Thus, the degrees $\deg_{\mathbb{C}}(\chi_G)$, resp. $\deg_{\mathbb{R}}(\chi_G)$, exist and one may ask for explicit formulae expressing their dependence on G. We shall consider the complex case first, when $n = mp$. According to Lemma 3.1, $\deg_{\mathbb{C}}(\chi_G)$ can be computed as the degree of the complex analytic map

$$\chi_G\colon \mathbf{Grass(p, U \oplus Y)} \to \mathbb{CP}^{\mathbf{mp}} \tag{3.8}$$

for generic G, i.e., for $G \in \mathbf{V}^{\mathbf{mp}}_{\mathbf{m,p}}$. Since $\deg(\chi_G)$ is the intersection number (3.5), and since the subspace

$$\mathbf{V}^{\mathbf{mp}}_{\mathbf{m,p}} \subset \Omega^{2}_{(\mathbf{mp})}\, \mathbf{Grass(m, m + p)}$$

is connected in the compact-open topology, $\deg(\chi_G)$ is *a priori* dependent only on m and p. First of all, as χ_G is proper and not constant, by elementary algebraic geometry, *image* χ_G is a closed subvariety of $\mathbb{CP}^{mp}$ of dimension mp and hence

$$\chi_G(\mathbf{Grass(p, U \oplus Y)}) = \mathbb{CP}^{\mathbf{mp}}. \tag{3.9a}$$

Furthermore, since

$$\chi_G(\sigma(\mathbf{U})) \subset \mathbb{CP}^{\mathbf{n}-1}_{\infty}$$

the same argument as above yields

$$\chi_G(\mathbb{C}^{\mathbf{mp}}) = \mathbb{C}^{\mathbf{n}}, \tag{3.9b}$$

i.e., for generic G and $mp = n$, over $\mathbb{C}$ one can place poles arbitrarily. That $\chi_G(\mathbb{C}^{\mathbf{mp}})$ is dense in $\mathbb{C}^{\mathbf{n}}$ was proved by Hermann–Martin in [29] using infinitesimal techniques.

One can also see that $\deg(\chi_G) > 0$ by noting that the Poincaré dual of the statement,

$$\chi^{-1}(\mathbb{CP}^{n-1}_\infty) = \boldsymbol{\sigma}(\mathbf{U})$$

as divisors, is the statement in cohomology

$$\chi^*(\omega) = \eta,$$

where ω, resp. η, is the generator of $\mathbf{H}^2(\mathbb{CP}^{mp}; \mathbb{Z})$, resp. $\mathbf{H}^2(\mathbf{Grass(p, U \oplus Y)}; \mathbb{Z})$. Recalling that the cohomology of $\mathbb{CP}^{mp}$ is a truncated polynomial ring in ω, $d = \deg_{\mathbb{C}}(\chi_G)$ is given by

$$\chi^*([\mathbf{vol}_{\mathbb{P}^{mp}}]) = \chi^*(\omega^{mp}) = \eta^{mp} = d[\mathbf{vol}_{\mathbf{Grass(p,\, U \oplus Y)}}]. \tag{3.10}$$

Since η is the class of a Kähler form on $\mathbf{Grass(p, U \oplus Y)}$, η^{mp} and hence d is nonzero. Moreover, the formula (3.10) show that d depends only on m and p. Indeed, from the Schubert calculus of enumerative geometry a formula for d is well-known. Explicitly,

$$d = \frac{1!2!\cdots(p-1)!(mp)!}{m!\cdots(m+p-1)!}, \tag{3.11}$$

which in the present context gives the number of complex feedback laws which assign a given set of (closed-loop) poles ([10]).

Remark. The connection with enumerative geometry is easier to see in terms of homology. That is, consider the equation

$$\chi_G(x) = y \tag{3.12}$$

for $x \in \mathbf{Grass(p, U \oplus Y)}$, $y \in \mathbb{CP}^{mp}$. Since y is an intersection

$$\{y\} = \bigcap_{i=1}^{mp} \mathbf{H}_i$$

of hyperplanes and since $\chi^{-1}(\mathbf{H}_i) = \boldsymbol{\sigma}(\mathbf{V}_i)$ for some m-plane $\mathbf{V}_i$, (3.12) may be rewritten as

$$\{x\} \in \bigcap_{i=1}^{mp} \boldsymbol{\sigma}(\mathbf{V}_i). \tag{3.12$'$}$$

Now, for $\mathbf{V}_i$ in general position, the p-planes x which satisfy (3.12)$'$ are precisely those p-planes which meet each of the mp m-planes $\mathbf{V}_i$ in $\mathbb{C}^{m+p}$ nontrivially. It is a classical problem in enumerative geometry to count, with multiplicity, the number of such p-planes. This was solved by Schubert in 1886 using his enumerative calculus which, however, relied on somewhat heuristic principles. In a partial answer to Hilbert's 15th Problem, Ehresmann, and Hodge, gave rigorous determinations of the homology and the algebraic (intersection) rings of Grassmannians which were sufficiently explicit to derive, for example, the formula (3.11)—see Kleiman's discussion of Hilbert's 15th Problem in *AMS Proc. of Symposia in Pure Mathematics*, Vol. 28, or Chapt. IV, §7 of [31].

Returning to (3.11), we make several special calculations. First, if $\min(m, p) = 1$, then $d = 1$ as it should be since, in this case, χ_G is linear. Second, in the case $m = p = 2$ considered by Willems and Hesselink ([54]), $d = 2$. Indeed, using standard elimination theory, Willems and Hesselink showed that the 4 by 4 system of real algebraic equations

$$\chi_G(K) = p$$

could be reduced, for generic G, to 3 linear equations and a single, nontrivial quadratic. Next, if $m = 2$ and $p = 3$, then $d = 5$. In particular, for real p there always exists a real solution K! Indeed,

Theorem 3.2 ([10]). *If* $\min(m, p) = 1$ *or* $\min(m, p) = 2$ *and* $\max(m, p) = 2^r - 1$, *then real solutions exist to the pole placement equations.*

To prove this theorem from the statements given above, one can either determine when d is odd or calculate when, in the mod 2 cohomology of the real Grassmannian, the cup-product η^{mp} is non-zero in analogy with the Kähler class argument sketched before. This latter calculation has been made by I. Bernstein in [3], yielding the conditions on m and p stated in the theorem.

Remark. The appearance of Mersenne numbers in this context was quite unexpected and one would naturally ask whether these conditions are necessary. Based on somewhat deeper calculations, it has been announced in [16] that

$$c_{m,p} \geq n \quad \text{implies pole-placement for generic } G \tag{3.13}$$

where $c_{m,p}$ is the Fox category of the Plücker imbedding

$$\mathbf{Grass(p, m + p)} \subset \mathbb{P}^N, \qquad N = \binom{m+p}{p} - 1.$$

Equation (3.13) implies Theorem 3.1 as well as several other partial results on this problem. It also implies, in fact, that when the conditions in Theorem 3.2 are not satisfied,

$$\deg_{\mathbb{R}}(\chi_G) = 0$$

for generic G, but whether or not χ_G fails to be surjective in this range remains open.

4. The Construction of the Classifying Spaces Rat(p, q; m)

Recall that $\mathbf{Rat(n; m)} \subset \boldsymbol{\Sigma}^{\mathbf{n}}_{\mathbf{m,m}}(\mathbb{R})$ is the subspace of real, rational *symmetric* $m \times m$ matrix-valued functions satisfying

$$G(\infty) = 0.$$

It is a theorem, proved in the appendix, that $\mathbf{\Sigma^n_{m,m}}(\mathbb{R})$ is naturally a smooth manifold of dimension $2nm$, whose underlying topology agrees with its compact-open topology when regarded as a space of real maps,

$$\mathbf{\Sigma^n_{m,m}}(\mathbb{R}) \subset \mathbf{\Sigma^n_{m,m}}(\mathbb{C}) \subset \mathbf{\Omega^2_{(n)}}(\mathbf{Grass(m, 2m)}).$$

Now **Rat(n; m)** is the fixed point set of the smooth involution

$$G \mapsto G^t$$

acting on $\mathbf{\Sigma^n_{m,m}}(\mathbb{R})$, and therefore

$$\mathbf{Rat(n; m)} = \cup \mathbf{F_k}$$

where $\mathbf{F_k}$ is a smooth manifold of dimension k, by Bochner's theorem ([5]). In this section, we shall prove that only the dimension $k = n(m + 1)$ occurs and that **Rat(n; m)** has $n + 1$ path components. More precisely, as in Section 1, each $\mathbf{G} \in \mathbf{Rat(n; m)}$ induces a map as

$$\mathbf{G}\colon \mathbb{RP}^1 \to \mathbf{LG(m, 2m)}$$

and expressing the Maslov index of **G** as

$$\mathrm{Ind}(\mathbf{G}) = p - q, \tag{4.1}$$

where $p + q = n$, we may decompose

$$\mathbf{Rat(n; m)} = \cup(\mathbf{Rat(p, q; m)} \tag{1.18}$$

in the compact-open topology. We shall show that each **Rat(p, q; m)** is a Euclidean connected, real analytic manifold, which is a classifying space for analytic families of transfer functions satisfying (4.1). The notation below is as in Definition 1.1.

Theorem 4.1. $\mathbf{G} \in \mathbf{Rat(p, q; m)}$ *if, and only if, G admits an internally symmetric realization with respect to* $I_{p,q}$. *Moreover, the natural map*

$$\mathbf{\Pi_{p,q}}\colon \widetilde{\mathbf{Rat}}\mathbf{(p, q; m)} \to \mathbf{Rat(p, q; m)}$$

is an analytic, principal **O(p, q)***-bundle with a smooth, connected base.*

PROOF. We shall analyze the map

$$\Pi\colon \widetilde{\mathbf{Rat}}\mathbf{(n; m)} \to \mathbf{Rat(n, m)}.$$

First of all, Π is surjective. For, if $G(s)$ is symmetric and defined over $\mathbb{R}$, Proposition 2.6 implies that $G(s)$ admits a minimal realization satisfying

$$\begin{aligned} I_{p,q}A &= A^t I_{p,q} \\ I_{p,q}B &= C^t \end{aligned} \tag{4.2}$$

for some signature matrix $I_{p,q}$. If $(A', B', C') \in \widetilde{\mathbf{Rat}}\mathbf{(p, q; m)}$ also realizes $G(s)$, there exists a unique $T \in \mathbf{GL(n, \mathbb{R})}$ satisfying

$$\alpha(T, (A, B, C)) = (A', B', C'). \tag{4.3}$$

Indeed, reachability of (A, B, C) implies $T \in \mathbf{O(p, q)}$ since the operator identity

$$TI_{p,q} = I_{p,q}(T^t)^{-1}$$

may be checked on span $(B, AB, \ldots, A^{n-1}B)$, and this is a trivial consequence of (4.2) and (4.3). Therefore, the action

$$\alpha\colon \mathbf{GL(n, k)} \times \tilde{\Sigma}^{\mathbf{n}}_{\mathbf{m,m}}(\mathbf{k}) \to \tilde{\Sigma}^{\mathbf{n}}_{\mathbf{m,m}}(\mathbf{k})$$

restricts, for $k = \mathbb{R}$ or $\mathbb{C}$, to an action

$$\alpha_1\colon \mathbf{O(p, q)} \times \widetilde{\mathbf{Rat}}(\mathbf{p, q; m}) \to \widetilde{\mathbf{Rat}}(\mathbf{p, q; m}). \qquad \square$$

Lemma 4.2. *The action α_1 is free and has Zariski closed orbits.*

PROOF. That α_1 is free follows (Proposition 2.2) from the freeness of α. Moreover, Lemma 2.2 of [20] asserts that the orbits of α are Zariski closed and, since $\mathbf{O(p, q)} \subset \mathbf{GL(n, k)}$ is an algebraic subgroup, the second statement follows. $\square$

By standard theorems on analytic group actions, it follows that

$$\pi_{\mathbf{p,q}}\colon \widetilde{\mathbf{Rat}}(\mathbf{p, q; m}) \to \widetilde{\mathbf{Rat}}(\mathbf{p, q; m})/\mathbf{O(p, q)} \tag{4.4}$$

defined via

$$\pi_{p,q}(A, B, C) = C(sI - A)^{-1}B$$

may be naturally regarded as an analytic principal $\mathbf{O(p, q)}$-bundle with an analytic manifold as base. Of course, (4.4) is a reduction of the structure group of the $\mathbf{GL(n, k)}$-bundle

$$\tilde{\Sigma}^{\mathbf{n}}_{\mathbf{m,m}}(\mathbf{k}) \to \Sigma^{\mathbf{n}}_{\mathbf{m,m}}(\mathbf{k})$$

over the subspace $\widetilde{\mathbf{Rat}}(\mathbf{p, q; m})/\mathbf{O(p, q)}$ (see Appendix).

Furthermore, the dimension of the base manifold is easily calculated as

$$\dim \widetilde{\mathbf{Rat}}(\mathbf{p, q}) - \dim \mathbf{O(p, q)} = p + q + nm = n(m + 1) \tag{4.5}$$

and therefore each of these $n + 1$ quotients is open in $\mathbf{Rat(n; m)}$, which is itself an analytic manifold of dimension $n(m + 1)$. This decomposition into submanifolds

$$\mathbf{Rat(n; m)} = \cup \widetilde{\mathbf{Rat}}(\mathbf{p, q; m})/\mathbf{O(p, q)} \tag{4.6}$$

shows that $\beta_0(\mathbf{Rat(n; m)}) \geq \mathbf{n + 1}$.

We wish to show that the submanifold $\widetilde{\mathbf{Rat}}(\mathbf{p, q; m})/\mathbf{O(p, q)}$ is connected and coincides with $\widetilde{\mathbf{Rat}}(\mathbf{p, q; m})$.

If $m = 1$, Brockett's Theorem asserts that

$$\beta_0(\mathbf{Rat(n)}) = \mathbf{n + 1},$$

so that $\widetilde{\mathbf{Rat}}(\mathbf{p}, \mathbf{q}; \mathbf{1})/\mathbf{O}(\mathbf{p}, \mathbf{q})$ is connected. Now suppose $m > 1$. To say (A, B, C) satisfying (1.10) is not reachable is to say that

$$\operatorname{rank}(B, AB, \ldots, A^{n-1}B) < n$$

which (since $m > 1$) imposes at least 2 independent algebraic constraints. On the other hand, for (A, B, C) satisfying (1.10), (A, B) is not reachable if, and only if, (A, C) is not observable. In particular

$$\mathbb{R}^{(n2+n+2nm)/2} \dot{-} \widetilde{\mathbf{Rat}}(\mathbf{n}, \mathbf{m})$$

has codimension at least 2, so that each $\widetilde{\mathbf{Rat}}(\mathbf{p}, \mathbf{q})$ and hence $\widetilde{\mathbf{Rat}}(\mathbf{p}, \mathbf{q}; \mathbf{m})/\mathbf{O}(\mathbf{p}, \mathbf{q})$ is connected.

We conclude the proof of Theorem 4.1 by examples showing

$$\mathbf{Rat}(\mathbf{p}, \mathbf{q}; \mathbf{m}) \cap \widetilde{\mathbf{Rat}}(\mathbf{p}, \mathbf{q}; \mathbf{m})/\mathbf{O}(\mathbf{p}, \mathbf{q}) \neq \varnothing$$

Consider the minimal, internally symmetric triple,

$$A = \operatorname{diag}(1, 2, \ldots, n), \qquad B = [b, 0, \ldots, 0], \qquad C = B^t I_{p,q}, \tag{4.7}$$

where $b^t = [1, \ldots, 1]$. An easy calculation yields

$$G(s) = C(sI - A)^{-1}B = \left(\sum_{i=1}^{p} \frac{1}{s-i} - \sum_{j=p+1}^{p+q} \frac{1}{s-j}\right)E_{11}. \tag{4.7$'$}$$

Lemma 4.3. $G(s)$, *as defined in* (4.7)$'$, *has Maslov index* $p - q$.

Proof. This can be computed directly for the case at hand but also follows from a general formula for the local contributions to $\operatorname{Ind}(G)$ regarded as an intersection number. If $s_0^- < s_0 < s_0^+$ are real points sufficiently close to an s_0 for which

$$\mathbf{G}(s_0) \in \boldsymbol{\sigma}(\mathbf{U}^*) \subset \mathbf{LG}(\mathbf{m}, \mathbf{U} \oplus \mathbf{U}^*)$$

then, according to Hörmander ([32], 3.3.4), the local intersection number at s_0 of $\mathbf{G}(\mathbb{RP}^1)$ with $\boldsymbol{\sigma}(\mathbf{U}^*)$ is given by

$$\operatorname{Ind}_{s_0}(\mathbf{G}) = (\operatorname{sgn} G(s_0^+) - \operatorname{sgn} G(s_0^-))/2 \tag{4.8}$$

under very general conditions on G. Explicitly, if $\boldsymbol{\sigma}(\mathbf{U})$ is the hypersurface in $\mathbf{LG}(\mathbf{m}, \mathbf{U} \oplus \mathbf{U}^*)$ defined by $\mathbf{U}$, then to say $\mathbf{G}(s) \in \boldsymbol{\sigma}(\mathbf{U})$ is to say $\det G(s) = 0$. Unless

$$\det G(s) \equiv 0$$

we can therefore assume that

$$\det G(s_0^-) \neq 0, \qquad \det G(s_0^+) \neq 0,$$

or that the Lagrangian planes

$$\mathbf{graph}\ \mathbf{G}(\mathbf{s}_0^-), \qquad \mathbf{graph}\ \mathbf{G}(\mathbf{s}_0^+)$$

are transverse to **U**. Now, the image of the interval $\mathbf{I} = [s_0^-, s_0^+]$ under **G** is a path in **LG(m, U ⊕ U*)** and Hörmander's calculation is valid whenever the path **G(I)** remains transverse to **U**, i.e., under the condition

$$\mathbf{G(I)} \subset \mathbf{LG(m, U \oplus U^*)} - \boldsymbol{\sigma}(\mathbf{U}).$$

Since the condition $\det G(s) = 0$, is algebraic in s one may choose s_0^-, s_0^+ so that

$$\det G(s) \neq 0 \quad \text{for all } s \in \mathbf{I},\ s \neq s_0.$$

That is, $\mathbf{G(I)} \cap \boldsymbol{\sigma}(\mathbf{U})$ is either empty or consists of the singleton $\{\mathbf{G}(s_0)\}$. Moreover, it follows from ([10], Section 2.14) that

$$\mathbf{G(I)} \cap \boldsymbol{\sigma}(\mathbf{U}) = \{\mathbf{G}(s_0)\}$$

if, and only if, the matrix valued function $G(s)$ has a zero at s_0—in addition to the pole. This phenomenon cannot, of course, occur when $G(s)$ is scalar. Now, for $G(s)$ defined in (4.7)′ we claim that for any $s_0 \in \{1, \ldots, n\}$, **I** can be chosen such that

$$\mathbf{G(I)} \cap \boldsymbol{\sigma}(\mathbf{U}) = \varnothing$$

and hence Hörmander's formula applies, yielding

$$\mathrm{Ind}(\mathbf{G}) = \sum_{G(s_0) = \infty} (\mathrm{sgn}\ G(s_0^+) - \mathrm{sgn}\ G(s_0^-))/2. \tag{4.8$'$}$$

Thus, assuming the claim, the statement in the lemma follows. As for the claim,

$$G(s) = g(s)E_{11}$$

so that the zeroes and poles of $G(s)$ are precisely those of $g(s)$. Since $g(s)$ is a scalar, a zero of $g(s)$ cannot coincide with a pole. Q.E.D.

The right hand side of Hörmander's formula (4.8)′ is closely related to an index which is in fact quite popular among control theorists (see, for example, [4] and [9]). Indeed, in the scalar case this is precisely the index originally considered by Cauchy [21]. In [4], the matrix Cauchy index for a rational, symmetric matrix valued function is defined as the sum, over real poles s_0, of a local index. Upon traversing such an s_0, the local index is calculated as the number of eigenvalues of $G(s)$ which change from $-\infty$ to $+\infty$ minus the number of eigenvalues which change from $+\infty$ to $-\infty$. If s_0 is not a zero of $G(s)$, then no eigenvalues of $G(s)$ can approach 0 so that this local index coincides with (4.8)′. On the other hand, if s_0 is a zero of $G(s)$ a negative (or positive) eigenvalue could deform through 0 to a positive (or negative) eigenvalue, in this case making a contribution to (4.8) but leaving the local index unchanged.

Thus, if the zeroes and poles of $G(s)$ do not coincide,

$$\textbf{Cauchy Ind(G)} = \textbf{Arnol'd–Maslov Ind(G)}.$$

We claim that this identity holds for all G, but as the remarks above show, one cannot use (4.8)′ to prove this statement. Here, we shall follow Arnol'd [1].

In general, composition with the Cayley transform induces a map

$$\mathbf{G}\colon \mathbf{S^1} \to \mathbf{U(m)}, \quad \text{defined via } \mathbf{G}(s) = (I - iG(s))(I + iG(s))^{-1} \tag{4.9}$$

and therefore leads to the invariant

$$[\mathbf{G}] \in \pi_1(\mathbf{U(m)}) \simeq \mathbf{Z}.$$

Now, $[\mathbf{G}]$ may be computed as

$$[\mathbf{G}] = \deg_{\mathbb{R}}(\det G(s)) \tag{4.10}$$

and, denoting by $\lambda_j(s)$ the algebraic functions of s satisfying

$$\det(\lambda I - G(s)) = 0$$

(3.22) yields

$$[\mathbf{G}] = \deg\left(\prod_{j=1}^{m} (1 - i\lambda_j(s))(1 + i\lambda_j(s))^{-1}\right). \tag{4.10$'$}$$

On the other hand, we claim

$$\deg\left(\prod_{j=1}^{m} (1 - i\lambda_j(s))(1 + i\lambda_j(s))^{-1}\right) = \sum_{G(s_0)=\infty} \textbf{Cauchy Ind}_{s_0}(G). \tag{4.11}$$

Now, use the left-hand side of (4.11) to calculate the degree of a product of algebraic functions

$$g_j(s) = (1 - i\lambda_j(s))(1 + i\lambda_j(s))^{-1}$$

which take values θ in $U(1)$ for s real. And, the degree in (4.11) is computed with respect to the base point $\theta = e^{i\pi}$ in $\mathbf{U(1)}$. Thus, the left-hand side is the sum of the degree of the algebraic functions $g_j(s)$. With these conventions, suppose the pole s_0 occurs also as a zero of $G(s)$, i.e. some $\lambda_j(s)$ vanishes at s_0 while some other eigen-function takes on infinite values. If we consider such a branch then on the one hand, as s goes through s_0, $\lambda_j(s)$ vanishes and hence makes no contribution to the degree of (its Cayley transform) $g_j(s)$. On the other hand, $\lambda_j(s)$ makes no contribution to the local Cauchy index of $G(s)$ at $s = s_0$, by definition.

Therefore,

$$[\mathbf{G}] = \textbf{Cauchy Ind}(G)$$

where the right-hand side is understood as the matrix Cauchy index, defined in [4].

Just as in the scalar case, one has a map

$$\pi\colon \mathbf{U(m)} \to \mathbf{LG(m, U \oplus U^*)} \simeq \mathbf{U(m)/O(m)},$$

and from the homotopy exact sequence of this fibration one obtains ([1], Corollary 3.4.3)

$$\pi_*\colon \pi_1(\mathbf{U(m)}) \simeq \pi_1(\mathbf{LG(m, U \oplus U^*)}).$$

Indeed,

$$\mathbf{Ind(G)} = \deg_{\mathbb{R}}(\det{}^2 G(s)) = \deg_{\mathbb{R}}(\det[(I - iG(s))(I + iG(s))^{-1}]).$$

This identity, together with (4.9), proves the fundamental identity

Theorem 4.4. Arnol'd-Maslov Ind(G) = Cauchy Ind(G).

Brockett has also been able to draw a rather interesting corollary in [8]. The (truncated) Hankel matrix $\mathscr{H}_G' = (h_{i+j-1})_{i,j=1}^n$ is constructed from the Laurent expansion

$$G(s) = \sum_{i=1}^{\infty} \frac{h_i}{s^i}$$

and is therefore a continuous symmetric matrix-valued function on **Rat(n; 1)** satisfying

$$\operatorname{rank}(\mathscr{H}'_G) = n, \quad \text{for } \mathbf{G} \in \mathbf{Rat(n; 1)}.$$

In particular, sgn $\mathscr{H}_G$ is also constant on the path components, and since the Hermite–Hurwitz Theorem

$$\operatorname{sgn}(\mathscr{H}'_G) = \mathbf{Cauchy\ Ind}(G) \tag{4.12}$$

holds for G defined by (4.7)$'$, this identity holds for all $\mathbf{G} \in \mathbf{Rat(n; 1)}$. By realization with parameters (see Appendix), the same proof for matrix-valued transfer functions yields a recent extension ([4], Theorem 3.1) of this beautiful and useful identity.

Corollary 4.5 (Anderson–Bitmead). *For any real symmetric $m \times m$ matrix-valued rational function $G(s)$, vanishing at ∞,*

$$\mathbf{Cauchy\ Ind(G)} = \operatorname{sgn}(\mathscr{H}'_{\mathbf{G}}).$$

One also obtains, by the same methods,

Corollary 4.6. *If $G(s)$ is the frequency response of an RLC network, then*

$$\mathbf{Ind(G)} = \#\,\mathbf{C} - \#\mathbf{L}.$$

Since the action of $\mathbf{O(p, q)}$ on $\widetilde{\mathbf{Rat}}\mathbf{(p, q; m)}$ is free, one can expect a formula for the element $\mathbf{T}_{\alpha\beta}(x) \in \mathbf{O(p, q)}$ which relates two local analytic sections $\gamma_\alpha = (A_\alpha, B_\alpha, C_\alpha)$, $\gamma_\beta = (A_\beta, B_\beta, C_\beta)$ on a trivializing open set $U_{\alpha\beta}$ of the bundle

$$\pi_{\mathbf{p},\mathbf{q}}: \widetilde{\mathbf{Rat}}(\mathbf{p}, \mathbf{q}; \mathbf{m}) \to \mathbf{Rat}(\mathbf{p}, \mathbf{q}; \mathbf{m}).$$

Indeed,

$$\mathbf{T}_{\alpha\beta}[B_\beta, A_\beta B_\beta, \ldots, A_\beta^{n-1} B_\beta] = [B_\alpha, A_\alpha B_\alpha, \ldots, A_\alpha^{n-1} B_\alpha]$$

defines $\mathbf{T}_{\alpha\beta}$ uniquely and gives rise to an analytic formula for $\mathbf{T}_{\alpha\beta}$, in light of reachability.

From this formula for the patching data $(\mathbf{T}_{\alpha\beta})$ it follows that one may construct a universal, internally symmetric family defined on the rank n vector bundle $\mathbf{V} \to \mathbf{Rat}(\mathbf{p}, \mathbf{q}; \mathbf{m})$ associated to the bundle $\pi_{\mathbf{p},\mathbf{q}}: \widetilde{\mathbf{Rat}}(\mathbf{p}, \mathbf{q}; \mathbf{m}) \to \mathbf{Rat}(\mathbf{p}, \mathbf{q}; \mathbf{m})$ by the contragredient of the representation $\mathbf{O}(\mathbf{p}, \mathbf{q}) \subset \mathbf{GL}(\mathbf{n}; \mathbb{R})$.

Corollary 4.7. $\mathbf{Rat}(\mathbf{p}, \mathbf{q}; \mathbf{m})$ *is a classifying space for continuous* (*smooth, or analytic*) *families of* $m \times m$ *symmetric transfer functions satisfying* (4.1).

Of course, the universal family is in general topologically nontrivial.

Corollary 4.8. *The* $\mathbf{O}(\mathbf{p}, \mathbf{q})$*-bundle*

$$\pi_{\mathbf{p},\mathbf{q}}: \widetilde{\mathbf{Rat}}(\mathbf{p}, \mathbf{q}; \mathbf{m}) \to \mathbf{Rat}(\mathbf{p}, \mathbf{q}; \mathbf{m})$$

is nontrivial whenever $m \geq 2$.

PROOF. If a global cross-section exists, we would have a decomposition.

$$\widetilde{\mathbf{Rat}}(\mathbf{p}, \mathbf{q}; \mathbf{m}) \simeq \mathbf{Rat}(\mathbf{p}, \mathbf{q}; \mathbf{m}) \times \mathbf{O}(\mathbf{p}, \mathbf{q}) \tag{4.13}$$

and hence $\widetilde{\mathbf{Rat}}(\mathbf{p}, \mathbf{q}; \mathbf{m})$ would need be disconnected, contrary to fact.

Q.E.D.

Remark. On the other hand, $\widetilde{\mathbf{Rat}}(\mathbf{p}, \mathbf{q}; \mathbf{1})$ is disconnected—one can show by brute force that

$$f(A, B) = \det(B, AB, \ldots, A^{n-1}B)$$

separates $\widetilde{\mathbf{Rat}}(\mathbf{p}, \mathbf{q}; \mathbf{1})$ into precisely 2 components. From such a calculation one could show by counting components in (4.13) that the universal bundles are trivial if, and only if, $m = 1$ and $\min(p, q) = 0$. The proof we shall offer in the next section, however, gives far more information about the topology of the base, $\mathbf{Rat}(\mathbf{p}, \mathbf{q})$, and this is our ultimate goal.

5. The Hermite–Hurwitz Bundles and the Topology of Spaces of Rational Functions

We shall first prove that, in the scalar case, the $\mathbf{O(p, q)}$-bundle

$$\pi_{\mathbf{p},\mathbf{q}}: \widetilde{\mathbf{Rat}}(\mathbf{p}, \mathbf{q}) \to \mathbf{Rat}(\mathbf{p}, \mathbf{q}) \tag{5.1}$$

introduced in Theorem 3.1 is trivial if, and only if, $p = 0$ or $q = 0$. This is, at first, interesting since the $\mathbf{GL(n, \mathbb{R})}$-bundle

$$\pi: \tilde{\Sigma}^{\mathbf{n}}_{\mathbf{1},\mathbf{1}}(\mathbb{R}) \to \Sigma^{\mathbf{n}}_{\mathbf{1},\mathbf{1}}(\mathbb{R}) = \mathbf{Rat(n)} \tag{5.2}$$

is trivial with cross-section given by

$$p(s)/q(s) \mapsto (A_q, e_1, (p_n, \ldots, p_1)), \tag{5.3}$$

where A_q is the companion matrix of q and $p(s) = s^n + p_n s^{n-1} + \cdots + p_1$. That the rational canonical form provides a regular cross-section for the geometric quotient of $\mathbf{GL(n, k)}$ acting on (reachable) pairs (A, b) consisting of an $n \times n$ matrix A, and a cyclic vector b was noted, for instance, in [45] and a straightforward extension of this observation leads to the cross-section (5.3) of the bundle (5.2). In particular, if $p = 0$ or $q = 0$, the bundle (5.1) is a reduction of the structure group of (5.2) to the maximal compact subgroup $\mathbf{O(n)}$, and is therefore trivial, although (5.3) is not a section of (5.1). Indeed, the construction of a cross-section of (5.1) is intimately related to the inverse spectral problem for Jacobi matrices.

On the one hand, the structure group of the bundle (5.1) can be reduced to the maximal compact subgroup $\mathbf{O(p)} \times \mathbf{O(q)}$ and, in light of the above, such a reduction is equivalent to the construction of a smooth, nondegenerate $n \times n$ symmetric matrix-valued function with signature $p - q$. And, the Hermite–Hurwitz Theorem provides a natural reduction in the form of the $n \times n$ (truncated) Hankel matrix $\mathscr{H}'_{\mathbf{G}}$, for $\mathbf{G} \in \mathbf{Rat(p, q)}$. On the other hand, one may adapt Lemma 2.4 to include the case of rings of C^∞ functions, yielding a criterion for the triviality of the bundle (5.1), viz., that $\mathscr{H}'_{\mathbf{G}}$ be conjugate over $\mathbf{C^\infty(Rat(p, q))}$ to the form $I_{p,q}$.

Now, $\mathscr{H}'_G$ defines two fibrations $\mathbf{V}_+\mathbf{(p, q)}$, $\mathbf{V}_-\mathbf{(p, q)}$ over $\mathbf{Rat(p, q)}$:

$$\mathbf{V}_\pm(\mathbf{p}, \mathbf{q}) \subset \mathbf{Rat}(\mathbf{p}, \mathbf{q}) \times \mathbb{R}^{\mathbf{p}+\mathbf{q}}$$

$$\mathbf{V}_\pm(\mathbf{p}, \mathbf{q}) = \{(\mathbf{G}, v): v \in \pm \text{ eigenspace of } \mathscr{H}'_{\mathbf{G}}\}.$$

In light of the Hermite–Hurwitz Theorem, $\mathbf{V}_\pm\mathbf{(p, q)}$ defines a real vector bundle over $\mathbf{Rat(p, q)}$, which we shall refer to as the positive (or negative) Hermite–Hurwitz bundle.

Lemma 5.1. *The* $\mathbf{O(p, q)}$*-bundle*

$$\pi_{\mathbf{p},\mathbf{q}}: \widetilde{\mathbf{Rat}}(\mathbf{p}, \mathbf{q}) \to \mathbf{Rat}(\mathbf{p}, \mathbf{q})$$

admits a cross-section if, and only if, the Hermite–Hurwitz bundles on $\mathbf{Rat(p, q)}$ *are trivial.*

PROOF. If the $\mathbf{O(p, q)}$-bundle is trivial, then so is the bundle $\mathbf{V}_\pm\mathbf{(p, q)}$ associated to the (reduced) $\mathbf{O(p)} \times \mathbf{O(q)}$-bundle. On the other hand, if $\mathbf{V}_\pm\mathbf{(p, q)} \to \mathbf{Rat(p, q)}$ is trivial, one may construct orthogonal frames for $\mathbf{V}_\pm\mathbf{(p, q)}$ which in turn piece together to define a cross-section of the (reduced) $\mathbf{O(p)} \times \mathbf{O(q)}$-bundle. Q.E.D.

Theorem 5.2. *The bundle*

$$\pi_{\mathbf{p},\mathbf{q}}\colon \widetilde{\mathbf{Rat}}\mathbf{(p, q)} \to \mathbf{Rat(p, q)}$$

admits a smooth cross-section if, and only if, $p = 0$ or $q = 0$. In other words, smooth canonical forms for internally symmetric realizations exist in the scalar case only if the Hankel matrix is either positive or negative definite.

Assuming the theorem, we may conclude

Corollary 5.3. *A smooth family $G(m; s)$ of scalar transfer functions, with fixed degree* $\deg(\mathbf{G})$, *on a smooth manifold M admits a smooth internally symmetric realization if, and only if, the Hermite–Hurwitz bundles*

$$\mathbf{f}_{\mathbf{G}}^{*}\mathbf{V}_\pm\mathbf{(p, q)} \to M$$

are trivial. If $|\mathrm{Ind}(\mathbf{G})| = \deg(\mathbf{G})$ *such a realization, depending smoothly on M, always exists.*

Here, of course,

$$\mathbf{f}_{\mathbf{G}}\colon M \to \mathbf{Rat(p, q)}$$

is the classifying map induced by the family $G(m; s)$, as in Corollary 4.6.

PROOF. We shall show that $\mathbf{V}_\pm\mathbf{(p, q)}$ is nontrivial when restricted to a suitable closed curve γ in $\mathbf{Rat(p, q)}$. By the homotopy invariance of vector bundles, this also shows that $\pi_1(\mathbf{Rat(p, q)})$ is nontrivial, as was announced in [15]. Indeed, consider the closed curve

$$G_\theta(s) = \frac{(\cos\theta)s + \sin\theta}{s^2 + 1} + \sum_{i=1}^{n-2} \frac{\varepsilon_i}{s - i}$$
$$\varepsilon_i = \begin{cases} 1 & \text{for } i = 1, \ldots, p - 1 \\ -1 & \text{for } i = p, \ldots, n - 2 \end{cases} \tag{5.4}$$

which lies in $\mathbf{Rat(p, q)}$. Now, $G_\theta(s)$ may be realized, for each θ, by

$$A_\theta = \begin{pmatrix} 0 & -1 & & & \\ 1 & 0 & & & \\ & & 1 & & \\ & & & 2 & \\ & & & & n-2 \end{pmatrix}, \qquad B_\theta = \begin{pmatrix} \sin\theta/2 \\ \cos\theta/2 \\ \varepsilon_1 \\ \vdots \\ \varepsilon_{n-2} \end{pmatrix} \tag{5.5a}$$

with

$$C_\theta^t = \begin{pmatrix} I_{1,1} & 0 \\ 0 & I_{p-1,q-1} \end{pmatrix} B_\theta. \tag{5.5b}$$

Thus, (5.5a-b) gives an internally symmetric realization of $G_\theta(s)$ which is smooth for $0 \le \theta \le \pi$ and for $\pi \le \theta \le 2\pi$. In particular, one may appeal to the identity (2.17) which leads to a patching description of $\mathbf{V}_\pm(\mathbf{p}, \mathbf{q})|_{S^1}$. That is, if $\mathbf{V}$ denotes the trivial $\mathbb{R}^n$-bundle over S^1 with fiber span$(\mathscr{H}'_{G(\theta)})$, then (2.17) and (5.5a-b) imply

$$\mathbf{V} = \mathbf{V}_1 \oplus \mathbf{W}_1 \oplus \mathbf{W}_2$$

with

(i) rank $\mathbf{V}_1 = 2$, rank $\mathbf{W}_1 = p - 1$, rank $\mathbf{W}_2 = q - 1$,
(ii) $\mathscr{H}'_{G(\theta)|\mathbf{W}_1} > 0$, $\mathscr{H}'_{G(\theta)|\mathbf{W}_2} < 0$, and
(iii) $\mathbf{W}_1$, $\mathbf{W}_2$ are trivial, with $\mathbf{V}_1 = (\mathbf{W}_1 \oplus \mathbf{W}_2)^\perp$.

Moreover, $\mathbf{V}_1$ is the bundle whose fiber over θ is span$(\mathscr{H}'_{g(\theta)})$ where

$$g(\theta) = \frac{(\cos\theta)s + \sin\theta}{s^2 + 1}. \tag{5.6}$$

From (5.6), one may compute

$$\mathscr{H}'_{g(\theta)} = \begin{pmatrix} \cos\theta & \sin\theta \\ \sin\theta & -\cos\theta \end{pmatrix} \tag{5.6$'$}$$

and it is easily verified that, for each $\theta \in \mathbf{S}^1$,

$$\mathbf{V}_+(\mathbf{1}, \mathbf{1})_{\{\theta\}} = \text{span}\{(\cos\theta/2, \sin\theta/2)\} \subset \mathbb{R}^2.$$

In particular, the classifying map induced by $\mathbf{V}_+(\mathbf{1}, \mathbf{1})|_{\mathbf{S}^1}$

$$\mathbf{f_g}: \mathbf{S}^1 \to \mathbb{RP}^1$$

has degree 1. Therefore, we have a decomposition

$$\mathbf{V}_1 \simeq \mathcal{O}(\mathbf{1}) \oplus \mathcal{O}(\mathbf{1}) \tag{5.7}$$

with $\mathscr{H}'_{G(\theta)}$ positive definite on one factor, and negative definite on the other. Since each vector bundle on $\mathbf{S}^1$ is a sum of line bundles, unique up to cancellations $\mathcal{O}(\mathbf{1}) \oplus \mathcal{O}(\mathbf{1}) \simeq \mathcal{O}^{(2)}$,

$$\mathbf{V} \simeq \mathbf{V}_+(\mathbf{p}, \mathbf{q})|_{\mathbf{S}^1} \oplus \mathbf{V}_-(\mathbf{p}, \mathbf{q})|_{\mathbf{S}^1} \tag{5.8}$$

where

$$\mathbf{V}_+(\mathbf{p}, \mathbf{q})|_{\mathbf{S}^1} \simeq \mathcal{O}(\mathbf{1}) \oplus \mathcal{O}^{(\mathbf{p}-\mathbf{1})}, \qquad \mathbf{V}_-(\mathbf{p}, \mathbf{q})|_{\mathbf{S}^1} \cong \mathcal{O}(\mathbf{1}) \oplus \mathcal{O}^{(\mathbf{q}-\mathbf{1})} \tag{5.9}$$

which proves the Theorem. Q.E.D.

Corollary 5.4 ([15]). **Rat(p, q)** *is simply-connected if, and only if,* $\min(p, q) = 0$.

PROOF. The proof of the theorem implies the necessity, while sufficiency follows from an observation made in [8]: to say $G(s)$ has Cauchy index n and degree n is to say

$$G(s) = \sum_{i=1}^{n} \frac{r_i}{s - \lambda_i}, \qquad \lambda_1 < \cdots < \lambda_n, \qquad r_i \in \mathbb{R}^+. \tag{5.10}$$

In particular, from (5.10) it follows that

$$\mathbf{Rat(n, 0)} \simeq \mathbb{R}^{2n} \simeq \mathbf{Rat(0, n)} \tag{5.11}$$

yielding in addition a second proof of the triviality of the $\mathbf{O(n)}$-bundle

$$\Pi_{n,0}\colon \widetilde{\mathbf{Rat}}(\mathbf{n}, \mathbf{0}) \to \mathbf{Rat(n, 0)}. \tag{5.12}$$

Q.E.D.

Since this result was announced, work by G. Segal has appeared ([50]) which allows us to be much more explicit about $\Pi_1(\mathbf{Rat(p, q)})$ and about the role of the Hermite–Hurwitz bundles. First of all, Segal constructs a rather neat change of coordinates. If

$$G(s) = n(s)/d(s), \qquad (n, d) = 1, \tag{5.13}$$

then one may form the complex polynomial

$$P(s) = n(s) + id(s). \tag{5.14}$$

It is quite clear from (5.13) that $P(s)$ has no real roots. More generally, the conjugates of the roots occurring in the lower half-plane $\mathbf{H}_-$ are distinct from those roots which occur in the upper half-plane $\mathbf{H}_+$. In particular, the numbers N_+, N_- of such roots are constant if G varies in a fixed component $\mathbf{Rat(p, q)}$ and, in fact,

$$N_+ = p, \qquad N_- = q. \tag{5.15}$$

Now, fix a diffeomorphism,

$$\Phi\colon \mathbf{H}_+ \simeq \mathbb{C}. \tag{5.16}$$

Conjugating $\mathbf{H}_-$ into $\mathbf{H}_+$ and applying Φ, one obtains for each G a divisor

$$\mathscr{D}_G = \Sigma n_i P_i - \Sigma m_i Q_i, \quad \deg(\mathscr{D}) = p - q$$

on $\mathbb{C}$ and hence a complex rational function $R_G(s)$, unique up to multiplication by $\mathbb{C}^*$. Suppose $p \geq q$, then by applying the Euclidean algorithm and noting that the remainder is arbitrary for $\mathbf{G} \in \mathbf{Rat(p, q)}$. Segal obtains

$$\mathbf{Rat(p, q)} \simeq \mathbb{R}^{2(p-q)} \times \mathbf{Rat}(\mathbf{q}; \mathbb{C}) \tag{5.17}$$

and, in particular, a conjecture of Brockett

$$\mathbf{Rat(p, q)} \simeq \mathbb{R}^{2(p-q)} \times \mathbf{Rat(q, q)}. \tag{5.18}$$

Since

$$\mathbf{Rat}(1; \mathbb{C}) \simeq \mathbb{C}^* \times \mathbb{C},$$

(5.17) yields

$$\mathbf{Rat}(\mathbf{n}, 1) \simeq \mathbf{Rat}(1, \mathbf{n}) \simeq \mathbf{S}^1 \times \mathbb{R}^{2\mathbf{n}+1} \tag{5.19}$$

which was derived by different techniques in [8]. More generally, one knows

$$\Pi_1(\mathbf{Rat}(\mathbf{q}; \mathbb{C})) \cong \mathbb{Z}, \qquad q \geq 1 \tag{5.20}$$

with a generator given by a loop moving a zero of $G_\theta(s)$ once around a pole of $G_0(s)$. This was proved originally by F. D. S. Jones and vastly extended by Segal who showed that the inclusion

$$\mathbf{Rat}(\mathbf{q}; \mathbb{C}) \subset \Omega^2_{(\mathbf{q})}(\mathbf{S}^2)$$

induces the isomorphism

$$\Pi_i(\mathbf{Rat}(\mathbf{q}; \mathbb{C})) \simeq \Pi_{i+2}(\mathbf{S}^2), \qquad i \leq q - 1.$$

In particular, (5.20) follows for $q > 1$ as well. In the appendix, we have sketched an algebraic geometric proof ([19]) of (5.20), see Theorem A.5. Combining (5.20) and the proof of Theorem 5.2, one obtains

Proposition 5.5. *If* $\min(p, q) > 0$, *then*

$$\mathrm{H}^1(\mathbf{Rat}(\mathbf{p}, \mathbf{q}); \mathbb{Z}_2) \simeq \mathbb{Z}_2$$

with the nonzero class given by the Stiefel–Whitney class

$$w_1(\mathbf{V}_+(\mathbf{p}, \mathbf{q})) = w_1(\mathbf{V}_-(\mathbf{p}, \mathbf{q})).$$

Conjecture 5.6. *The* mod(2) *cohomology ring of* **Rat(p, q)** *is generated by the Stiefel–Whitney classes of the Hermite–Hurwitz bundles.*

Remark. On the one hand, the contact with the Hermite–Hurwitz Theorem and with the homotopy groups of 2-sphere makes it quite interesting to study the algebraic topology of the various spaces **Rat(p, q)**. On the other hand, the topology of **Rat(p, q)** has a great influence on the system theory of scalar input–output systems. In the first place, analysis of the existence of smooth or continuous canonical forms for internally symmetric realizations of transfer functions gives rise to primary obstructions in the cohomology of **Rat(p, q)**. Furthermore, it has long been of interest to know whether one can define a vector field on **Rat(p, q)** having a sink as its only critical point. In Section 6, we shall apply these results to an analysis of the equilibrium manifolds of certain vector fields on **Rat(p, q)**.

For the sake of completeness, we record the following results.

Corollary 5.7. *The* **O(p, q)**-*bundles*

$$\widetilde{\mathbf{Rat}}(\mathbf{p}, \mathbf{q}; \mathbf{m}) \to \mathbf{Rat}(\mathbf{p}, \mathbf{q}; \mathbf{m})$$

admit a continuous cross-section if, and only if, $m = 1$ *and* $\min(p, q) = 0$.

This result should be contrasted with the algebraic situation, e.g., Corollary 2.5. More explicitly,

Corollary 5.7′. *Continuous canonical forms for minimal, internally symmetric realizations exist only in the scalar input–output case and only when the Hankel matrix is positive or negative semi-definite.*

We shall now turn to the explicit construction of a cross-section of the **O(n)**-bundle

$$\pi_{\mathbf{n}, \mathbf{0}}\colon \widetilde{\mathbf{Rat}}(\mathbf{n}, \mathbf{0}) \to \mathbf{Rat}(\mathbf{n}, \mathbf{0}) \tag{5.12}$$

or, what is the same up to diffeomorphism, the bundle

$$\pi_{\mathbf{0}, \mathbf{n}}\colon \widetilde{\mathbf{Rat}}(\mathbf{0}, \mathbf{n}) \to \mathbf{Rat}(\mathbf{0}, \mathbf{n}).$$

In one setting, the inverse problem is to construct a self-adjoint matrix A and a cyclic vector b for A from the spectral data:

$$G(s) = \langle b, (sI - A)^{-1} b \rangle. \tag{5.21}$$

Developing the partial fraction expansion (5.11) in a continued fraction expansion

$$G(s) = \cfrac{1}{s - a_{nn} - \cfrac{a_{n-1,n}^2}{s - a_{n-1,n-1}} \cdots - \cfrac{a_{1,2}^2}{s - a_{1,1}}} \tag{5.22}$$

leads to the Jacobi matrix

$$A = \begin{pmatrix} a_{11} & a_{12} & & 0 \\ a_{12} & a_{22} & & \\ & & \ddots & a_{n-1,n} \\ 0 & & a_{n-1,n} & a_{n,n} \end{pmatrix} \tag{5.23}$$

which has e_n as a cyclic vector. In the RC setting, it was noted by Cauer ([22]) that the triple (A, ce_n, ce_n^t), where

$$c = \sqrt{\sum_{i=1}^{n} r_i}$$

realizes the frequency response $G(s)$, while in spectral theory this inverse method is also quite classical ([52], [53], esp. pp. 507ff.). In any case, it is easy to show that the submanifold Jac(n), consisting of minimal Jacobi triples (A, ce_n, ce_n^t), is transverse to the fibers of the map (5.12). Since

$$\Pi_{\mathbf{n}, \mathbf{0}}\colon \mathbf{Jac}(\mathbf{n}) \to \mathbf{Rat}(\mathbf{n}, \mathbf{0}) \tag{5.24}$$

is surjective by construction and since

$$\mathbf{Rat(n, 0)} \simeq \mathbb{R}^{2n} \tag{5.12}$$

is simply connected, the covering map (5.24) is injective. Therefore, Jac(n) constitutes an analytic cross-section of the (trivial) $\mathbf{O(n)}$-bundle (5.12).

6. An Application to the Study of Globally Convergent Vector Fields on Spaces of Systems

Motivated especially by questions of adaptive control and identification, one would like to develop a qualitative theory of smooth vector fields on **Rat(n; m)** in so far as one is able in the absence of compactness. For example, since

$$\mathbf{Rat(n)} \simeq \mathbb{R}^{2n} - \mathscr{V}(\mathbf{Res})$$

as in the appendix, each component **Rat(p, q)** is parallelizable and, in particular, admits vector fields having no zeros. However, the question which is of interest for certain applications is whether **Rat(p, q)** admits a vector field having a sink as its only critical point. More generally, we shall investigate the possibility that a submanifold $N \subset$ **Rat(p, q)** is locally and globally attracting for some smooth vector field, $X \in$ Vect(**Rat(p, q)**).

Theorem 6.1. *If* $X \in$ Vect(**Rat(p, q)**) *and* $N \subset$ **Rat(p, q)** *is a nonempty compact orientable submanifold, which is locally and globally attracting for* X, *then either* $\min(p, q) = 0$ *or* $\min(p, q) = 1$. *Furthermore, the only* 2 *possibilities are*

(a) *N is a point if, and only if,* $\min(p, q) = 0$, *and*
(b) $N \simeq S^1$ *if, and only if,* $\min(p, q) = 1$.

PROOF. It is of course obvious from

$$\mathbf{Rat(n, 0)} \simeq \mathbf{Rat(0, n)} \simeq \mathbb{R}^{2n} \tag{5.17}$$

and

$$\mathbf{Rat(n - 1, 1)} \simeq \mathbf{Rat(1, n - 1)} \simeq \mathbf{S^1} \times \mathbb{R}^{2n-1}, \tag{5.19}$$

that on each of these components a vector field possessing the appropriate ω-limit may be constructed. The proof of (a) is rather straightforward and is based on the following weak result.

Lemma 6.2. *If* $X \in$ Vect(M) *and* $p \in M$ *is a global sink for* X, *then* $\pi_1(M) = \{0\}$.

PROOF. Choose a coordinate neighborhood U for x in M and let $[\gamma] \in \pi_1(M; x)$. For each $y \in \gamma$, there exists a time $T \geq 0$ such that

$$\Phi_T(y) \in U, \tag{6.1}$$

where Φ_T is the flow induced by X. And, by continuity, all points $y' \in \gamma$ sufficiently close to y also satisfy (6.1). By a standard compactness argument, there exists a finite time T such that

$$\Phi_T(\gamma) \subset U \tag{6.1$'$}$$

and therefore

$$O = (\Phi_T)_*[\gamma] = [\gamma] \in \pi_1(M; x). \tag{6.2}$$

Q.E.D.

In particular, (a) follows from Corollary 5.4. We remark that the proof of Lemma 6.2 also implies that all the homotopy groups of M must vanish.

Statement (b), however, lies much deeper. If $N \subset \mathbf{Rat(p, q)}$ is globally attracting for X, then it is not hard to show that N must be connected. Moreover, we claim:

Lemma 6.3. *If* $\min(p, q) > 0$, *then N has the rational homology of* $\mathbf{S^1}$.

PROOF. As above choose a tubular neighborhood U of N in $\mathbf{Rat(p, q)}$. Since for $x \in N$,

$$\pi_i(U; x) \simeq \pi_i(N; x), \quad \text{for } i \geq 1$$

the same proof as above shows that

$$\pi_i(N; x) \simeq (\mathbf{Rat(p, q)}; x), \quad \text{for } i \geq 1 \tag{6.3}$$

so that, in particular, by the Whitehead Theorem

$$H_i(N; \mathbb{Q}) \simeq H_i(\mathbf{Rat(p, q)}; \mathbb{Q}), \quad \text{for } i \geq 0. \tag{6.4}$$

By virtue of (5.17), we may rewrite (6.4) as

$$H_i(N; \mathbb{Q}) \simeq H_i(\mathbf{Rat(r}; \mathbb{C}); \mathbb{Q}), \quad i \geq 0 \tag{6.4$'$}$$

where $r = \min(p, q)$. Now, Segal has also defined inclusions ([50])

$$\mathbf{Rat(r}; \mathbb{C}) \subset \mathbf{Rat(r + 1}; \mathbb{C}) \subset \dots.$$

He proves that these inclusions induce injections in homology and has also shown that the homology eventually stabilizes. In particular, for rational coefficients,

$$H_i(\mathbf{Rat(r + 1}; \mathbb{C}); \mathbb{Q}) = H_i(\mathbf{Rat(r}; \mathbb{C}); \mathbb{Q}) \oplus V_r^i$$

where $V_r^i = \{0\}$, for $r \gg 0$. On the other hand, since

$$\mathbf{Rat(r}, \mathbb{C}) \subset \mathbf{\Omega^2_{(r)}(S^2)}$$

induces isomorphisms

$$\pi_i(\mathbf{Rat(r}, \mathbb{C})) \simeq \pi_i(\mathbf{\Omega^2_{(r)}(S^2)}), \quad \text{for } i \leq r - 1,$$

it follows from Whitehead's Theorem that, for $r \gg 0$,

$$H_*(\mathbf{Rat(r; \mathbb{C})}; \mathbb{Q}) \simeq H_*(\Omega^2_{(r)}(\mathbf{S}^2); \mathbb{Q}). \tag{6.5}$$

But, it is well-known that

$$H_*(\Omega^2_{(r)}(\mathbf{S}^2); \mathbb{Q}) \simeq H_*(\mathbf{S}^1; \mathbb{Q})$$

where the generator $\mathbf{S}^1 \subset \Omega^2_{(r)}(\mathbf{S}^2)$ may be represented by the Hopf map. These results imply, therefore, that

$$H_*(\mathbf{Rat(r; \mathbb{C})}; \mathbb{Q}) \simeq H_*(\mathbf{S}^1; \mathbb{Q})$$

for $r \gg 0$. Noting, however, that

$$\mathbf{Rat(1; \mathbb{C})} \simeq \mathbb{C}^* \times \mathbb{C},$$

in harmony with Brockett's calculation (5.19), one obtains

$$H_*(\mathbf{Rat(r; \mathbb{C})}: \mathbb{Q}) \simeq H_*(\mathbf{S}^1; \mathbb{Q})$$

whenever $r = \min(p, q) \geq 1$. Q.E.D.

By Stokes' Theorem, $\dim(N) = 1$, leaving only the possibility

$$\mathbf{N} \simeq \mathbf{S}^1 \tag{6.6}$$

which can occur if $\min(p, q) = 1$. To finish the proof of the theorem, suppose $r = \min(p, q) \geq 2$. In this case, we may compute

$$\pi_2(\mathbf{Rat(p, q)}) = \begin{cases} \mathbb{Z} & \text{if } r = 2 \\ \mathbb{Z}_2 & \text{if } r > 2. \end{cases}$$

That is, Segal's Theorems assert that the composite map

$$\pi_2(\mathbf{Rat(p, q)}) \simeq \pi_2(\mathbf{Rat(r; \mathbb{C})} \to \pi_4(\mathbf{S}^2) \simeq \mathbb{Z}_2$$

is an isomorphism when $r > 2$ and a surjection when $r = 2$. One may calculate $\pi_2(\mathbf{Rat(2, 2)})$ by explicit methods which we shall not exhibit here. In any case. Segal's work shows that

$$\pi_2(\mathbf{Rat(p, q)}) \neq \{0\}$$

if $\min(p, q) > 1$, while (6.3) and (6.6) yield

$$\pi_2(\mathbf{Rat(p, q)}) = \{0\}$$

contrary to fact. Therefore, only the identity $\min(p, q) = 1$ is tenable.

Q.E.D.

Remark 1. In general, stronger topological statements are possible. To begin with the proof of Lemma 6.2 also shows, as we have noted in the proof of Lemma 6.3, that if M is a smooth manifold admitting $X \in \text{Vect}(M)$ which possesses a (unique) locally and globally attracting equilibrium $p \in M$, then

$$\pi_i(M) = \{0\} \quad \text{for } i \geq 1$$

and hence M is contractible, by Whitehead's Theorem. Moreover, following Milnor's proof of the generalized Reeb Theorem ([41], Theorem 1′), we see that if M is paracompact then M *is homeomorphic to* $\mathbb{R}^n$. More generally, F. W. Wilson has shown that if $N \subset M$ is locally and uniformally globally attracting for $X \in \text{Vect}(M)$, then M is diffeomorphic to an open tubular neighborhood of N—recovering Milnor's result when $N = \{pt\}$. Although these beautiful results will no doubt find an application in the topological theory of identification and adaptive control, for the present purposes it is sufficient to make our calculations at the level of homotopy.

Remark 2. There are quite a few interesting flows and, more generally, group actions on the manifolds **Rat(p, q; m)** and it is fair to say that a significant portion of the open problems in linear system theory amounts to an analysis of these flows and actions. One class of flows which arise are the various scaling operations on transfer functions ([11]), for example, an additive change in the time scale produces a flow

$$Ce^{At}B \to Ce^{A(t+T)}B \tag{6.7}$$

on the space of fundamental solutions to the system (1.9). As P. S. Krishnaprasad has observed ([38]), on $\mathscr{H}^n_{1,1}(\mathbb{R})$ the flow (6.7) takes the form

$$\frac{dH_i}{dt} = H_{i+1} \tag*{(6.7)$'$}$$

and on **Rat(n, 0)** we have, in the coordinates (5.10), the equation for the Toda lattice ([42])

$$\begin{aligned} \frac{d\lambda_i}{dt} &= 0 \\ \frac{dr_i}{dt} &= \lambda_i r_i \end{aligned} \tag*{(6.7)$''$}$$

by virtue of the correspondence (A.3)′. This, of course, is not suprising in light of the fact that the change of coordinates

$$\pi_{\mathbf{n},\mathbf{0}}\colon \mathbf{Jac(n)} \to \mathbb{R}^{\mathbf{2n}} \tag{5.24}$$

plays a crucial role in Moser's explicit integration of the Toda lattice.

Appendix. The Construction of Manifolds of Linear Systems, Realization with Parameters

In this appendix, we shall derive several theorems, some of which are new, concerning the ambient moduli spaces

$$\mathbf{Rat(n; m)} \subset \Sigma^n_{m,m}(\mathbb{R}) \subset \Sigma^n_{m,m}(\mathbb{C})$$

introduced in Proposition 2.2. First of all, for $k = \mathbb{R}$ or $\mathbb{C}$ the space $\Sigma^n_{m,p}(k)$ is a smooth quasiaffine variety defined over k. This follows from geometric invariant theory, or from the familiar alternate description of $\Sigma^n_{m,p}(k)$, viz. on the set $\mathscr{H}^n_{m,p}(k)$ of $p \times m$ block Hankel matrices, having rank n. We shall give a new proof of the following basic result.

Theorem A.1 ([23]). *$\mathscr{H}^n_{m,p}(k)$ is a smooth manifold.*

PROOF. If $G(s)$ has rank n, then the finite Hankel matrix $H_G = (H_{i+j-1})^{n+1}_{i,j=1}$ has rank n and determines $G(s)$. Now, if $\mathscr{M}_n(k)$ denotes the smooth variety of $(n+1)p$ by $(n+1)m$ matrices with rank n, then $\mathscr{H}^n_{m,p}(k)$ is evidently the fixed point set for an algebraic, hence smooth, action of the finite group

$$\mathscr{G} = \mathbb{Z}_2^{(2)} \times \cdots \times \mathbb{Z}_n^{(2)} \times \mathbb{Z}_{n+1}$$

on $\mathscr{M}_n(k)$. By Bochner's Theorem ([5]), since $\mathscr{G}$ is compact, $\mathscr{H}^n_{m,p}(k)$ is a smooth submanifold of $\mathscr{M}_n(k)$. Q.E.D.

Corollary A.2. *$\mathscr{H}^n_{m,p}(k)$ is a smooth quasi-affine variety.*

PROOF. $\mathscr{M}_n(\mathbb{C})$ is quasi-affine, i.e., is a Zariski open subset of an affine variety and $\mathscr{H}^n_{m,p}(\mathbb{C})$ is Zariski closed in $\mathscr{M}_n(\mathbb{C})$. Of course, since $\mathscr{H}^n_{m,p}(\mathbb{C})$ is nonsingular as a smooth manifold, it is a smooth variety. Similar remarks now apply to the set of real points, $\mathscr{H}^n_{m,p}(\mathbb{R})$. Q.E.D.

The second remark is concerned with various alternative descriptions of the spaces $\Sigma^n_{m,p}(\mathbb{C})$, these are

(Aa) as the quasi-projective variety $\Sigma^n_{m,p}(\mathbb{C})$ ([13]) of algebraic maps

$$\mathbf{G}\colon \mathbb{CP}^1 \to \mathbf{Grass}(\mathbf{m}, \mathbf{U} \oplus \mathbf{Y})$$

of degree n, satisfying the base point condition

$$\mathbf{G}(\infty) = [\mathbf{U}];$$

(Ab) ([20], [28]) as the quotient variety $\tilde{\Sigma}^n_{m,p}(\mathbb{C})/\mathbf{GL}(\mathbf{n}, \mathbb{C})$ for the action α defined in (2.6);

(Ac) ([23]) as the nonsingular, quasi-affine variety $\mathscr{H}^n_{m,p}(\mathbb{C})$ of complex $p \times m$ block Hankel matrices, having rank n.

And, there are natural maps

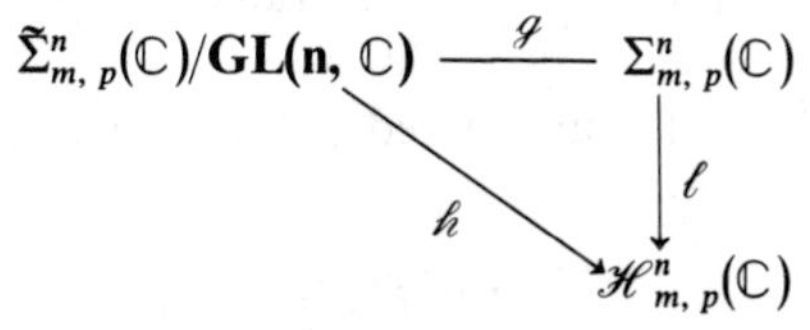

$$\begin{array}{ccc} \tilde{\Sigma}^n_{m,p}(\mathbb{C})/\mathbf{GL}(\mathbf{n}, \mathbb{C}) & \xrightarrow{\;g\;} & \Sigma^n_{m,p}(\mathbb{C}) \\ & \searrow^{h} & \downarrow \ell \\ & & \mathscr{H}^n_{m,p}(\mathbb{C}) \end{array} \tag{A.1}$$

defined via

$$\begin{aligned} \mathscr{g}([A, B, C])(s_0) &= \text{graph } C(s_0 I - A)^{-1}B \\ \mathscr{h}([A, B, C]) &= (CA^{i+j-2}B) \qquad\qquad (A.1)' \\ \ell(G(s)) &= H_G. \end{aligned}$$

It is a basic fact, which we shall refer to as "realization with parameters," that (A.1) is a commutative diagram of biregular equivalences.

This has been proven ([14] Proposition D, [27] Corollary 2.5.7) for the map $\mathscr{h}$ and therefore we need only prove that either $\mathscr{g}$ or ℓ is biregular. It suffices to check that ℓ is a morphism on an affine cover (U_α) of $\Sigma^n_{m,p}(\mathbb{C})$, and we denote the coordinate ring of U_α by R_α. Consider the morphism

$$\eta_\alpha : U_\alpha \times \mathbb{P}^1 \to \mathbf{Grass(m, U \oplus Y)}$$

defined by

$$(\mathbf{G}, s) \mapsto \mathbf{G}(s), \quad \text{for } \mathbf{G} \in U_\alpha,$$

and note that η_α satisfies the base variety condition

$$\eta_\alpha(U_\alpha \times \{\infty\}) = [\mathbf{U}].$$

In other words, η_α defines (after deleting $\boldsymbol{\sigma}(\mathbf{Y})$ from $\mathbf{Grass(m, U \oplus Y)}$) a transfer function over the Noetherian ring R_α ([13], Section 3). From realization theory over R_α, one obtains

(i) a state module $\mathscr{Q}$,
(ii) an endomorphism $\mathscr{A}: \mathscr{Q} \to \mathscr{Q}$, and
(iii) homomorphisms $\mathscr{B}: R_\alpha^{(m)} \to \mathscr{Q}$, $\mathscr{C}: \mathscr{Q} \to R_\alpha^{(p)}$

which realize η_α. In particular, forming the Hankel matrix leads to a morphism

$$\ell_\alpha : U_\alpha \to \mathscr{H}^n_{m,p}(\mathbb{C})$$

which agrees with ℓ on U_α.

Now, let $V_1 \subset \Sigma^n_{m,p}(\mathbb{C})$ denote the quasi-projective variety of maps having distinct poles, i.e., simple intersection with the Schubert hypersurface $\boldsymbol{\sigma}(\mathbf{Y})$, and let $V_2 \subset \mathscr{H}^n_{m,p}(\mathbb{C})$ denote the quasi-projective variety whose points correspond to systems having distinct eigenvalues. Thus, $\ell: V_1 \to V_2$ is a bijection. Moreover, if $\mathbf{G} \in V_1$ then as a transfer function G admits a partial fraction decomposition

$$G(s) = \sum_{i=1}^{n} \frac{R_i}{s - \lambda_i}, \quad \text{rank } R_i = 1 \qquad (A.2)$$

with λ_i distinct. Now the unordered n-tuple $(\lambda_1, \ldots, \lambda_n)$ is a regular morphism

$$V_1 \to \mathbb{A}^\mathbf{n}/\mathbf{S_n} \simeq \mathbb{A}^\mathbf{n}$$

since it is the divisor on $\mathbb{P}^1$ defined by the intersection of the rational curve $\mathbf{G}(\mathbb{P}^1)$ with the Schubert hypersurface $\boldsymbol{\sigma}(\mathbf{Y})$. Therefore, (A.2) gives rise to regular coordinates, viz., the unordered tuple $(\lambda_i, R_i)_{i=1}^n$ on V_1. In these coordinates, ℓ may be computed via

$$\ell = (\ell_1, \ldots, \ell_n; (\lambda_1, \ldots, \lambda_n)) \tag{A.3}$$

where

$$\ell_j : (\lambda_i, R_i)_{i=1}^n \mapsto \sum_{i=1}^n \lambda_i^{j-1} R_i = CA^{j-1}B. \tag{A.3$'$}$$

For, knowledge of $\mathbf{P_G}(t) = \prod_{i=1}^n (t - \lambda_i)$ enables one to compute ℓ_{n+1} via the Cayley–Hamilton Theorem. Conversely, a Vandermonde argument shows that (A.3)$'$ may be inverted and hence ℓ defines a biregular equivalence between V_1 and V_2. Now consider the graph of the rational correspondence ℓ^{-1},

$$gr(\ell^{-1}) \subset \mathscr{H}^n_{m,p}(\mathbb{C}) \times \Sigma^n_{m,p}(\mathbb{C}).$$

We claim that ℓ^{-1} is extendable as a continuous map of $\mathscr{H}^n_{m,p}(\mathbb{C})$ to $\Sigma^n_{m,p}(\mathbb{C})$, endowed with the compact-open topology. Assuming the claim, we may appeal to the Riemann Extension Theorem to deduce that ℓ^{-1} is a morphism. Equivalently,

$$\text{proj}_1 : gr(\ell^{-1}) \to \mathscr{H}^n_{m,p}(\mathbb{C})$$

is a bijective birational morphism and since $\mathscr{H}^n_{m,p}(\mathbb{C})$ is smooth Zariski's Main Theorem implies that ℓ^{-1} is a morphism.

As for the claim, to say ℓ^{-1} is continuous is to say $\mathscr{g}$ is continuous, but this is clear since the map

$$(A, B, C, s) \mapsto \text{graph } C(sI - A)^{-1}B$$

is jointly continuous.

There are several corollaries which follow immediately from the "realization with parameters" lemma.

Theorem A.3. *$\Sigma^n_{m,p}(\mathbb{C})$ is a nonsingular, irreducible quasi-affine variety of dimension $n(m+p)$, with Picard group a quotient of* $\mathbf{GL(n,}\ \mathbb{C}) \simeq \mathbb{Z}$. *Indeed,*

$$\text{Pic}(\Sigma^n_{m,p}(\mathbb{C})) \simeq \begin{cases} \mathbb{Z} & \text{if } \min(m, p) > 1 \\ \mathbf{(0)} & \text{otherwise} \end{cases}$$

with generator a complete obstruction to the existence of an algebraic canonical form for realizations.

Proof. Since $\Sigma^n_{m,p}(\mathbb{C}) \simeq \mathscr{H}^n_{m,p}(\mathbb{C})$, $\Sigma^n_{m,p}(\mathbb{C})$ is a smooth irreducible quasi-affine variety. And, since $\Sigma^n_{m,p}(\mathbb{C})$ is the base of an algebraic $\mathbf{GL(n,}\ \mathbb{C})$-bundle with total space $\tilde{\Sigma}^n_{m,p}(\mathbb{C})$ a Zariski open subspace of $\mathbb{C}^{\mathbf{N}}$, $N = n^2 + nm + np$, the remaining statements follow—except for an explicit determination of the Picard group (see [20], Section 5) as a particular quotient of $\mathbb{Z}$. Q.E.D.

Corollary A.4. $\Sigma^n_{m,p}(\mathbb{C}) \subset \mathbf{\Omega}^2_{(n)}\,\mathbf{Grass(m, n+p)}$ *is a connected manifold in the compact-open topology.*

ALTERNATE PROOF. The total space $\tilde{\Sigma}^n_{m,p}(\mathbb{C})$ of the fiber bundle described in $A(b)$ is the complement of an affine algebraic set in $\mathbb{C}^N$, $N = n^2 + nm + np$, viz. the union of the algebraic sets of nonreachable systems with the algebraic set of unobservable systems. In particular, the algebraic set

$$\mathbb{C}^N - \tilde{\Sigma}^n_{m,p}(\mathbb{C}) = X$$

has real codimension at least 2 and therefore $\Sigma^n_{m,p}(\mathbb{C})$ is connected. Thus, the base space $\tilde{\Sigma}^n_{m,p}(\mathbb{C})$ must also be connected. Q.E.D.

When $n = 1$, $\Sigma^n_{1,1}(\mathbb{C}) \simeq \mathbf{Rat(n}, \mathbb{C})$ and much more is known about the inclusion,

$$\mathbf{Rat(n}, \mathbb{C}) \subset \mathbf{\Omega}^2_{(n)}(\mathbf{S}^2). \tag{A.4}$$

Indeed, G. Segal has shown ([50]) that (A.4) induces a homotopy equivalence, up to degree n. In particular, for $n > 1$

$$\pi_1(\mathbf{Rat(n}, \mathbb{C})) \simeq \pi_3(\mathbf{S}^2) \simeq \mathbb{Z} \tag{A.5}$$

while (A.5) is easy to verify when $n = 1$ as well. The assertion (A.5) was previously proved by F. D. S. Jones using a configuration space argument. An elementary algebraic geometric proof of (A.5) based on the diffeomorphism

$$\mathbf{Rat(n}, \mathbb{C}) \simeq \mathcal{H}^n_{1,1}(\mathbb{C})$$

has been given in [19] and we shall give a sketch of this proof below.

Theorem A.5([50]). $\pi_1(\mathbf{Rat(n}, \mathbb{C})) \simeq \mathbb{Z}$, *with a generator given by any loop based at $G(s)$ which moves a zero (of $G(s)$) once around a pole (of $G(s)$).*

PROOF ([19]). The proof is based on a rather simple but neat corollary to a certain Lefschetz-type hyperplane theorem (see [26]).

Lemma A.6. *If* $\mathbf{V}$ *is an irreducible normal hypersurface in* $\mathbb{C}^N$, *then* $\pi_1(\mathbb{C}^N - \mathbf{V}) \simeq \mathbb{Z}$ *with generator any loop which encircles* $\mathbf{V}$ *once.*

PROOF. One cannot apply the lemma directly to

$$\mathbf{Rat(n}, \mathbb{C}) = \mathbb{C}^{2n} - \mathscr{V}(\mathbf{Res})$$

since the zero set $\mathscr{V}(\mathbf{Res})$ of the *resultant*

$$\mathbf{Res}(G) = \text{resultant}(\text{num}(G), \text{den}(G))$$

is not normal, i.e.,

$$\dim \mathscr{V}(\mathbf{Res})_{\mathbf{sing}} = \dim \mathscr{V}(\mathbf{Res}) - 1 = 2n - 2.$$

However, it is rather straightforward to check ([19]) that

$$\mathscr{H}^n_{1,1}(\mathbb{C}) = \mathbb{C}^{2n} - \mathscr{V}(\mathbf{det})$$

is the complement of a normal hypersurface. That is, if

$$\mathbf{det}(\mathscr{H}_G) \equiv \det(\mathscr{H}'_G) = \det(\mathscr{H}_{i+j-1})^n_{i,j=1},$$

then to say $\mathscr{H}_G$ is a singular point on $\mathscr{V}(\mathbf{det})$ is to say, in particular,

$$\frac{\partial \mathscr{H}_G}{\partial \mathscr{H}_{2n-1}} = 0. \tag{A.6}$$

The light of (A.6), classical arguments involving Hankel matrices imply that $\operatorname{rank}(\mathscr{H}_G) \leq n - 2$ and therefore

$$\dim \mathscr{V}(\mathbf{det})_{\mathbf{sing}} = \dim \mathscr{V}(\det) - 2. \qquad \text{Q.E.D.}$$

Our third remark is that, since the maps and spaces in (A.1) are all defined over $\mathbb{R}$, the lemma on "realization with parameters" holds over $\mathbb{R}$ as well and one can expect corollaries similar to those obtained in the complex case. For example, we can obtain a proof of Glover's theorem:

Proposition A.7 ([25]). *$\mathscr{H}^n_{m,p}(\mathbb{R})$ is connected provided* $\max(m, p) > 1$.

PROOF. We shall show $\Sigma^n_{m,p}(\mathbb{R})$ is connected in this range of m and p. Indeed, the open dense subspace $U^n_{m,p}$ consisting of those $\mathbf{G}$ with distinct poles is connected. To see this, note that any $\mathbf{G_0} \in U^n_{m,p}$ can be expressed in a partial fraction expansion

$$G_0(s) = \sum_{i=1}^n \frac{R_i^0}{s - \lambda_i}, \quad \operatorname{rank} R_i^0 = 1.$$

Moreover, a standard divisor argument shows that $\mathbf{G_0}$ may be deformed within $U^n_{m,p}$ to $\mathbf{G_1}$, having all real poles:

$$G_1(s) = \sum_{i=1}^n \frac{R_i^1}{s - r_i}, \qquad r_1 < \cdots < r_n$$

where each R_i' is a real $p \times m$ rank 1 matrix. If $\max(m, p) > 1$, then the space $\mathscr{R}$ of such residues is connected and hence $U^n_{m,p}$ is connected. Q.E.D.

Remark. If $m = p = 1$, then the space of residues is $\mathbb{R} - \{0\}$ and therefore there exist at least

$$P_2(n) = \#\text{ of partitions of } n \text{ into 2 parts} = n + 1$$

path components in $U^n_{1,1}$. R. W. Brockett has shown that, in fact, $\mathbf{Rat(n)} = \Sigma^n_{1,1}(\mathbb{R})$ has precisely $n + 1$ path components

$$\mathbf{Rat(n)} = \dot{\bigcup_{\substack{p+q=n \\ p,q \geq 0}}} \mathbf{Rat(p, q)}$$

distinguished by $\deg_{\mathbb{R}}(\mathbf{G}) = p - q$.

That $\Sigma_{m,p}^n(\mathbb{R})$ is a smooth manifold of dimension $n(m+p)$ was also proved independently by Byrnes and Hurt, and by Hazewinkel and Kalman, by proving that

$$\tilde{\Sigma}_{m,p}^n(\mathbb{R}) \to \tilde{\Sigma}_{m,p}^n(\mathbb{R})/\mathbf{GL}(\mathbf{n}, \mathbb{R}) \tag{A.7}$$

is a principal $\mathbf{GL}(\mathbf{n}, \mathbb{R})$-bundle.

This is shown in [28] by rather nice, quite explicit constructions, inspired to a great extent by the underlying control theory. Here it is also shown that the principal bundle (A.7) is algebraically nontrivial (over $\mathbb{R}$) and an admirable attempt at deriving explicit equations, which would exhibit this quotient as a quasiprojective variety imbedded in a $\mathbb{P}^N$, is made. In [20] (announced in [13]), geometric invariant theory was invoked to obtain the conclusion that the quotient in (A.7) exists and is a quasiprojective variety. Aside from the fact that the general techniques in [43] specialize to the barehands construction given in [28], the rather amazing and perhaps most striking side of this application of geometric invariant theory is that, for the action of $\mathbf{GL}(\mathbf{n}, \mathbf{k})$ on pairs (A, B), the reachable systems (A, B), form precisely the set of properly stable points (for the standard character linearization) in the sense of Mumford [43]. By combining our proof of Theorem A.1 with the following lemma, one can give an elementary proof of the 2 principal results in [28]: that $\Sigma_{m,p}^n(\mathbb{R})$ is smooth and that (A.7) is a nontrivial bundle whenever $\min(m, p) > 1$—which is the real analogue of Theorem A.3.

Lemma A.8. $\tilde{\Sigma}_{m,p}^n(\mathbb{R})$ *is connected just in case* $\min(m, p) > 1$.

Proof. If $\min(m, p) > 1$ then both the real algebraic set of unreachable systems and the real algebraic set of unobservable systems have codimension ≥ 2 in $\mathbb{R}^N$, $N = n^2 + nm + np$. On the other hand if, say, $m = 1$ then the algebraic set of systems has exactly 2 components. A similar remark applies to the case $p = 1$. Q.E.D.

Corollary A.9. *The bundle* (A.7) *is nontrivial whenever* $\min(m, p) > 1$. *In other words, continuous canonical forms for state-space realizations do not exist if* $\min(m, p) > 1$.

Proof. To say (A.7) is trivial is to say

$$\tilde{\Sigma}_{m,p}^n(\mathbb{R}) \simeq \Sigma_{m,p}^n(\mathbb{R}) \times \mathbf{GL}(\mathbf{n}, \mathbb{R}).$$

If $\min(m, p) > 1$, however, $\tilde{\Sigma}_{m,p}^n(\mathbb{R})$ and (hence) $\Sigma_{m,p}^n(\mathbb{R})$ are connected, but $\mathbf{GL}(\mathbf{n}, \mathbb{R})$ is disconnected—a contradiction. Q.E.D.

As a final application of "realization with parameters," and to explain the origin of its name, we shall give an alternate proof that condition (2.18), viz.

$$\operatorname{rank} \mathscr{H}_G \bmod M = n(M) \equiv n, \quad M \in \operatorname{Max}(R)$$

ensures the existence of a controllable, observable realization over R with free state module. Explicitly, (2.18) implies that one has a morphism

$$\mathscr{H}: \mathbb{A}^{\mathbf{N}} \to \mathscr{H}^{n}_{m,p}(\mathbb{C}).$$

Composing with $\hbar^{-1}$, one may pull back the universal family on $\tilde{\Sigma}^{n}_{m,p}(\mathbb{C})/\mathbf{GL(n, \mathbb{C})}$ along $\hbar^{-1} \circ \mathscr{H}$ obtaining

(i) a vector bundle $\mathbf{E} \to \mathbb{A}^{\mathbf{N}}$,
(ii) an endomorphism $A: \mathbf{E} \to \mathbf{E}$,
(iii) a homomorphism $B: \mathbb{A}^{\mathbf{N}} \times \mathbb{C}^{\mathbf{m}} \to \mathbf{E}$,
(iv) a homomorphism $C: \mathbf{E} \to \mathbb{A}^{\mathbf{N}} \times \mathbb{C}^{\mathbf{P}}$,

satisfying (2.15)′, (2.16)′ by construction. Finally, by the Quillen–Suslin Theorem, one knows that $\mathscr{H}_G$ is realized by a triple of matrices over R, satisfying (2.15)′ and (2.16)′.

Remark. The existence of such a realization is, in fact, equivalent to the Quillen–Suslin Theorem, for suppose a finitely generated, projective module Q over R is given. That is, suppose

$$Q \oplus Q' = R^{(N)}, \quad \text{for some } Q'.$$

Setting $B = proj_1$, $A = O$, $C = inclusion\ of\ Q\ in\ R^{(N)}$, one constructs a Hankel matrix $\mathscr{H}_G = (CA^{i+j-2}B)$ satisfying (2.18) and for which the existence of a controllable, observable realization by a triple of matrices defined over R implies that Q is free.

References

[1] V. I. Arnol'd, Characteristic class entering in quantization conditions, *Funct. Anal. Appl.* **1** (1967), 1–13.

[2] H. Bass, Quadratic modules over polynomial rings, in *Contributions to Algebra*, H. Bass, P. Cassidy, and J. Kovavic, Eds., Academic Press, New York, NY, 1977, 1–23.

[3] I. Berstein, On the Lusternick–Schnirelmann category of grassmannians, *Math. Proc. Camb. Phil. Soc.* **79** (1976), 129–134.

[4] R. R. Bitmead and B. D. O. Anderson, The matrix Cauchy index: Properties and applications, *SIAM J. Appl. Math.* **33** (1977), 655–672.

[5] S. Bochner, Compact groups of differentiable transformations, *Ann. Math.* **46** (1945), 372–381.

[6] R. K. Brayton and J. K. Moser, A theory of nonlinear networks—I, *Quart. Appl. Math.* **XXII** (1964), 1–33.

[7] R. W. Brockett, *Finite Dimensional Linear Systems*, John Wiley, New York, NY, 1970.

[8] —, Some geometric questions in the theory of linear systems, *IEEE Trans. Aut. Control* **21** (1976), 449–455.

[9] —, Lie algebras and rational functions: Some control theoretic questions, in *Lie Theories and Their Applications*, W. Rossmann, Ed., Queens University, Kingston, ON, Canada, 1978, 268–280.

[10] R. W. Brockett and C. I. Byrnes, Multivariable Nyquist criteria, root loci, and pole placement: A geometric viewpoint, *IEEE Trans. Aut. Control* **26** (1981), 271–284.
[11] R. W. Brockett and P. S. Krishnaprasad, A scaling theory for linear systems, *IEEE Trans. Aut. Control* **25** (1980), 197–207.
[12] R. Bumby, E. D. Sontag, H. J. Sussmann, and W. Vasconcelos, Remarks on the pole-shifting problem over rings, to appear in *J. Pure and Appl. Algebra*, **20** (1981).
[13] C. I. Byrnes, The moduli space for linear dynamical systems, in *Geometric Control Theory*, C. Martin and R. Hermann, Eds., Math. Sci. Press, Brookline, MA, 1977, 229–276.
[14] —, On the control of certain deterministic infinite-dimensional systems of algebro-geometric techniques, *Amer. J. Math.* **100** (1978), 1333–1381.
[15] —, On certain problems of arithmetic arising in the realization of linear systems with symmetries, *Astérisque* **75–76** (1980), 57–65.
[16] —, Recent results on output feedback problems, in *Proc. 19th IEEE Conf. on Decision and Control*, Albuquerque, NM, 1980, 663–664.
[17] C. I. Byrnes and T. E. Duncan, A note on the topology of spaces of Hamiltonian transfer functions, in *Algebraic and Geometric Methods in Linear System Theory*, AMS Lectures in Applied Math. Vol. 18, 1980, Providence, RI, 7–26.
[18] C. I. Byrnes and P. L. Falb, Applications of algebraic geometry in system theory, *Amer. J. Math.* **101** (1979), 337–363.
[19] C. I. Byrnes and B. K. Ghosh, On the fundamental group of spaces of coprime polynomials, to appear.
[20] C. I. Byrnes and N. E. Hurt, On the moduli of linear dynamical systems, *Adv. in Math. Studies in Analysis* **4** (1979), 83–122. (Translated in *Math. Methods in System Theory*, MIR, Moscow, 1979.)
[21] A Cauchy, Calcul des indices des fonctions, *J. L'École Polytechnique* (1835), 196–229.
[22] W. Cauer, *Theorie der Linearen Wechselstromschaltungen*, Akademie Verlag, Berlin, 1954.
[23] J. M. C. Clark, The consistent selection of local coordinates in linear system identification, in *Proc. JACC*, Purdue, 1976, 576–580.
[24] T. E. Duncan, An algebro-geometric approach to estimation and stochastic control for linear pure delay time systems, in *Lecture Notes in Control and Information Sci.*, Vol. 16, Springer-Verlag, New York, NY, 1979, 332–343.
[25] K. Glover, Some geometrical properties of linear systems with implications in identification, in *Proc. IFAC*, Boston, MA, 1975.
[26] H. Hamm and Lê Dũng Trang, Un théorème de Zariski du type Lefschetz, *Ann. Scient. Éc. Norm. Sup.* **6** (1973), 317–355.
[27] M. Hazewinkel, Moduli and canonical forms for linear dynamical systems, III: The algebraic-geometric case," in *Geometric Control Theory*, C. Martin and R. Hermann, Eds., Math. Sci. Press, Brookline, MA, 1977, 291–336.
[28] M. Hazewinkel and R. E. Kalman, On invariants, canonical forms, and moduli for linear constant, finite-dimensional dynamical systems, *Lecture Notes Econ.-Math. System Theory* **131** (1976), 48–60.
[29] R. Hermann and C. F. Martin, Applications of algebraic geometry to systems theory—Part I, *IEEE Trans. Aut. Control* **22** (1977), 19–25.
[30] ——, Applications of algebraic geometry to systems theory: The McMillan degree and Kronecker indices of transfer functions as topological and holomorphic system invariants, *SIAM J. Control Optim.* **16** (1978), 743–755.
[31] W. V. D. Hodge and D. Pedoe, *Methods of Algebraic Geometry*, Vol. II, Cambridge Univ. Press, Cambridge, England, 1952.
[32] L. Hörmander, Fourier integral operators, I, *Acta Math.* **127** (1971), 79–183.

[33] A. Hurwitz, Über die Bedingungen unter Welchen eine Gleichung nur Wurzeln mit Negativen Reelen Theilen Besitzt, *Math. Ann.* **46** (1895), 273–284.
[34] R. E. Kalman, Mathematical description of linear dynamical systems, *SIAM J. Control* **1** (1963), 128–151.
[35] E. W. Kamen, On an algebraic theory of systems defined by convolution operators, *Math. Systems Theory* **9** (1975), 57–74.
[36] E. W. Kamen, An operator of linear functional differential equations, *J. Diff. Eqs.* **27** (1978), 274–297.
[37] M. Knebusch, Grothendieck- und Wittringe von Nichtausgearteten Symmetrischen Bilinearformen, *Sitz. Heidelberg Akad. Wiss.* (1969/70), 93–157.
[38] P. S. Krishnaprasad, Geometry of minimal systems and the identification problem, Ph.D. Thesis, Harvard Univ., Cambridge, MA, 1977.
[39] L. Kronecker, Zur Theorie der Elimination einer Variablen aus Zwei Algebraischen Gleichnung, *Trans. Roy. Press. Acad.* (1881) (*Collected Works*, Vol. 2).
[40] S. Lefschetz, *Applications of Algebraic Topology*, Springer-Verlag, New York, NY, 1975.
[41] J. W. Milnor, Differential topology, in *Lectures in Modern Mathematics*, Vol. II, T. L. Saaty, Ed., J. Wiley, New York, NY, 1964.
[42] J. Moser, Finitely many mass points on a line under the influence of an exponential potential—An integrable system, *Lecture Notes in Physics* **38** (1975), 467–497.
[43] D. Mumford, *Geometric Invariant Theory*, Springer-Verlag, Berlin, 1965.
[44] —, *Lectures on Curves on an Algebraic Surface*, Annals of Math. Studies No. 59, Princeton Univ. Press, Princeton, NJ, 1966.
[45] D. Mumford and K. Suominen, Introduction to the theory of moduli, in *Algebraic Geometry*, F. Oort, Ed., Oslo, Norway, 1970, 171–222.
[46] M. Ojanguren, Formes quadratiques sur les algèbres de polynomes, *C.R. Acad. Sc. Paris*, Sér. A **287** (1978), 695–698.
[47] S. Parimala, Failure of a quadratic analogue of Serre's conjecture, *Amer. J. Math.* **100** (1978), 913–924.
[48] I. Postlethwaite and A. G. J. MacFarlane, *A Complex Variable Approach to the Analysis of Linear Multivariable Feedback Systems*, Lecture Notes in Control and Inf. Sciences, Vol. 12, Springer-Verlag, Berlin, 1979.
[49] Y. Rouchaleau, B. F. Wyman, and R. E. Kalman, Algebraic structure of linear dynamical systems III. Realization theory over a commutative ring, in *Proc. Nat. Acad. Sci.* **69** (1972), 3404–3406.
[50] G. Segal, The topology of spaces of rational functions, *Acta Math.* **143** (1979), 39–72.
[51] E. D. Sontag, On split realizations of response maps, *Inf. and Control* **37** (1978), 23–33.
[52] T. J. Stieltjes, Recherches sur les fractions continues, *Ann. Fac. Sci. Toulouse* **8** (1894), 1–122; **9** (1895), 1–47.
[53] M. H. Stone, *Linear Transformations in Hilbert Space*, Coll. Pub. 15, Amer. Math. Soc., Providence, RI, 1932.
[54] J. C. Willems and W. H. Hesselink, Generic properties of the pole-placement problem, in *Proc. 1978 IFAC*, Helsinki, Finland.
[55] F. W. Wilson, The structure of the level surfaces of a Lyapunov function, *J.O.D.E.* **3** (1967), 323–329.
[56] D. C. Youla, The synthesis of networks containing lumped and distributed elements, *Networks and Switching Theory* **11** (1968), 289–343.
[57] D. C. Youla and P. Tissi, N-port synthesis via reactance extraction—Part I, *IEEE Intern. Convention Record* (Mar. 1966), 183–205.

Operations Research and Discrete Applied Mathematics

D. Kleitman*

Operations research is an area of great potential for growth. Managing of any kind of system consists to a large extent of diagnosing, recommending, and supervising change. As more and more aspects of society become computerized, such management requires understanding what programs and machines can do and how one changes them to do whatever is best in a new situation.

The need for people who can perform this function and can implement their decisions has already created a growing demand for mathematicians, not only as programmers, but as the management personnel who can direct programmers.

A typical problem in operations research is finding a "best" course of action among many alternatives. The problem and its solution have the following aspects.

1. The situation, in which there are various limitations what can be done, and various merits and demerits of allowable courses of action.
2. A model, in which variables are introduced, the limitations reduced to constraint equations or inequalities, and the benefits described by an objective function to be maximized.
3. Techniques, for finding solutions to the "mathematical program" defined by the model, which hopefully yield recommended actions.

All of this is in direct analogy with what happens in describing behavior of physical or chemical systems. There the models typically lead to differential equations and the techniques of interest involve solution of differential

* Department of Mathematics, Massachusetts Institute of Technology, R 2-334, Cambridge, MA 02139.

or integro-differential equations. In this latter context the major effect of the computer revolution has been to give mathematicians new tools and as a result to expand their ability to solve equations enormously.

In optimization something different has happened. Computers have enormously increased the importance of the problems, and have made a number of methods practical, but have not led to practical general methods of solution, as we shall see.

There are some problems that can be handled very neatly by our techniques. These include, for example, those that can be modelled by constraints that are linear in their variables with an objective function likewise linear. For these there is a procedure found by Dantzig and others known as the simplex algorithm which has had great success in handling even very large problems (hundreds of variables, thousands of constraints according to some experts).

Many more problems can be formulated as "integer linear programs" in which the variables are constrained to be integers as well as obeying linear constraints, with a linear objective function. Before discussing the status of such problems let us look at an example.

The Traveling Salesman Problem

Suppose there are n points $v_1, \ldots, v_n$ with a cost c_{ij} associated with moving from v_i to v_j, for each (i, j). Suppose further that we seek a circuit C passing once through all the vertices having minimum cost. This is the "traveling salesman" or minimum cost Hamiltonian circuit problem.

This problem is actually of considerable practical interest. Not only is it faced by the proverbial salesman, but also by routers of school buses, and beer delivery trucks, to name only two. These latter face this problem for each bus or truck, but that is only a component of their problem which involves trying to get by with a minimum number of vehicles. As a result they sometimes use techniques that involve solution of many 20 to 50 node traveling salesman problems, often in an iterative fashion to improve solutions.

The problem can be written as an integer LP in several ways. The easiest of these involves very many constraints, most of which are irrelevant to any problem. We set $x_{ij} = 1$ if our circuit moves from v_i to v_j, and $x_{ij} = 0$ if otherwise. We then minimize $\sum c_{ij} x_{ij}$ subject to constraints that assure us of having a circuit, namely

$$\sum_j x_{ij} = \sum_j x_{ji} = 1 \quad \text{for all } i \tag{A}$$

for every proper subset A of vertices

$$\sum_{i \in A,\, j \notin A} x_{ij} \geq 1. \tag{B}$$

We will illustrate our discussion below with reference to this problem. Attacks on IP's have proceeded in three directions.

First, theorists have attempted to develop general methods for handling them. Among these are "relaxation" techniques in which one ignores some of the constraints of the problem, solves the resultant problem and then uses techniques for reintroducing what has been violated. In the traveling salesman problem, for example, only a few of the constraints (B) are ever relevant to a solution of the problem, and these can be introduced as they are violated. "Cutting plane methods" that are ways of reintroducing integrality conditions and "Lagrangian relaxation" techniques have been extensively studied, and have some advocates as practical methods.

It is difficult to assess how successful these efforts to solve general integer programs have been. Those involved with them suggest that their methods are useful, particularly on problems of moderate size, up to 30 or 40 variables, say. I cannot tell whether or not these claims are inflated to attract consulting business or not.

A second approach to such problems has been explored by computer scientists interested in complexity theory. When computers first appeared on the horizon, some enthusiasts believed that they could do almost anything, solve any problem and entirely replace mathematicians. The problem of what problems they could solve and what they couldn't given unlimited time was then studied. It soon became clear, however, that solvability in unlimited time is a concept of no practical utility. In recent years much effort has gone into investigating more practical measures of complexity. Perhaps the most successful of these has been the notion of NP-completeness which we will now describe.

A class of problems is said to be "in P" if each problem in it can be solved in a number of steps that is bounded by a polynomial of finite degree characteristic of the class, with variable given by the number of symbols necessary to describe the problem.

A class of problems is "in NP" if, in effect, one can verify whether or not a proposed solution actually is one in a number of steps bounded by a polynomial as above.

(Nobody now knows if NP and P define the same classes.) A class C is said to be NP-complete if every problem in any class in NP can be reduced in a polynomially bounded number of steps to a problem in C.

It is commonly believed that P is not the same as NP, in which case there is no polynomially bounded algorithm for NP complete problems. The lack of such an algorithm is generally interpreted as implying the impossibility of a practical general algorithm for the problem.

One of the triumphs of this viewpoint has been the fact that all sorts of problems have been proven to be NP-complete, and hence probably not practically generally solvable. In particular, integer linear programming is an NP complete class of problems, as is the class defined by the question, "Does a 'traveling salesman' problem have a solution with cost less than c?" for all cost functions and c's.

In other words, the thrust of this approach has been to prove that in a sense these problems cannot be solved. It has been shown that even finding a solution to traveling salesman problems with approximately optimal cost is *NP*-complete. It should of course be noted that this sense here refers to "worst case" behavior. A class may be *NP*-complete because of one subclass while a subclass containing all problems of practical interest may be easily solvable.

A third approach to these problems has been pursued by people who are actually faced with them. A number of algorithms have been developed that solve them more or less.

A typical example is Shen Lin's algorithm of the traveling salesman problem. This algorithm proceeds from a random starting circuit, through local improvement switches of arcs until a local optimum is attained. Then another start is made. After all local optima have been achieved a number of times the best is declared the solution. One can vary the extent of the local optimization procedure to give faster or alternatively, better solutions. Lin claims that his algorithm usually gives the optimum answer, and can be used easily with more than a hundred nodes.

Other heuristic methods start with solutions to other problems that are not too difficult to find, and make sequences of small improvements on these to obtain good solutions.

In most practical cases nobody knows how good the best of these algorithms is, since nobody knows the best answer. Despite this fact most people who use traveling salesman problems practically consider it a solved problem. They can obtain a packaged algorithm, in fact several of them, that solve the problem reasonably well within their desired timescale.

Thus there is the following anomalous situation in this field. If we consider the travelling salesman problem, one group of practitioners works on general approaches with some optimism; another is convinced that the problem is basically unsolvable, while *OR* practitioners consider it solved.

Now everyone may well be right. The worst case conclusion may well not have any implication in regard to the practical problems that arise in this field. These latter may well have additional structure that make the "general" or "heuristic" approaches work well.

It is clear, however, that there is much work to be done in recognizing the additional structure inherent in real classes of problems, and extracting their implications. It is unlikely that general solutions will solve all *IP* problems, but we will be able to generate useful solutions to many *IP* problems if one pays attention to their structure.

In a way this is good news for mathematicians in this field. If a simple general solution to all these problems was obtained, or if no progress at all could be made, we would soon be phased out of the problem. As it is, there is plenty of good mathematics left.

In order to illustrate the flavor of work in this area, we briefly describe three recent results. Two of these are rather minor and are presented primarily because I worked on them.

The Hajian Algorithm for Linear Programming

Dantzig's simplex algorithm for linear programs has long been the centerpiece of operations research technique. In it one moves along edges from vertex to vertex of the polytope defined by the constraints always improving the value of the objective function until one reaches the solution.

In practice this algorithm works very rapidly and has been used extensively on very large problems.

In it, one can use different rules for deciding where to move from one vertex to the next when there are several directions in which the objective function improves. With some of the rules, including the most commonly used in practice, one can construct examples of problems in which the number of steps is exponential in the size of the problem.

Thus the simplex algorithm, which appears to be remarkably useful, was not proven to be necessarily polynomial in the size of the problem. It was not known whether *LP* problems were in *P*.

Hajian showed that *LP* is in *P* by showing that an algorithm that had been considered by others could be used to solve linear programs in polynomial time.

We use it on the problem of finding a point obeying all the constraints of an *LP* or proving there is no such point. The algorithm proceeds by iterating the following step. Suppose we know that any solution to the *LP* must be in some n-sphere. Then we can ask: is the center of the sphere a good solution? If it is we are done—we have a solution point. Otherwise we can find a constraint the center violates, and more than half the sphere must violate it also. If we enclose the part of the sphere that obeys it in a suitable n-ellipsoid, and make a scale change to make that into an n-sphere, we know that all solutions lie in this new n-sphere, which has smaller volume than the original one. The constraints are still hyperplanes if they were so before. With repetition of this step, the allowed volume goes down exponentially in the number of repetitions if the center is never a solution. Hajian noticed that there is a maximum and minimum volume of an open finite polytope determined by given constraints, so that one can start with all solutions in some large n-sphere, and after a polynomial number of iterations, declare that the volume has become too small and there are no solutions if none have been found.

Practical use of this algorithm is not inconceivable, but not very promising as a competitor to the simplex algorithm for large problems. A number of investigators have been attempting to make it practical, with some encouragement but not ultimate success as yet.

Testing for Failure of a Line

Suppose messages are sent that are relayed along a line through n stations; station j has a probability p_j of failure, and the probability of two simultaneous failures is small enough to be ignored. One can test the line by choosing

k and finding out if the signal is getting to station k. One can raise the question: if one discovers that the line has failed somewhere, what is the shortest sequence of tests that will determine the failed station?

This problem has a neat solution. Each test divides the stations into two groups: those up to the one tested in which failure of the test would indicate error; and those beyond it, in which successful communication in the test indicates failure.

If we assign a first digit 0 to the stations in the first set and 1 to those in the second, assigning further digits according to location of the second test, etc., a test plan corresponds to an alphabetic prefix-free binary code. There is a well known algorithm for constructing an optimal code of this kind, known as the Hu–Tucker algorithm.

Frank Hwang has raised the question, if the p_j's are chosen from a given set of n numbers, what ordering of these gives the highest expected number of tests. Mike Saks and I were able to show that arranging the p's such that (for n odd, with a similar zigzag for even n)

$$p_1 \leq p_3 \leq \cdots \leq p_n \leq p_{n-1} \leq \cdots \leq p_4 \leq p_2$$

requires the most testing.

The problem arises in bounding the behavior of the Hu–Tucker algorithm. This result is surprisingly difficult to prove. Try it if you doubt me.

Blocking L Pentaminoes

One place progress in these problems may come is from new methods. But of course new methods are hard to come by. One way to look for these is to find what appear to be intractable simple problems and slug away at them.

To this end a simple problem that had not achieved an answer was that of finding the minimum density of a set of unit squares (with corners at integro-coordinate points) on the plane whose occupation prevents placement of any five block L shaped region (3×3 with a 2×2 removed from one corner).

Fernando Lasaga and I have been able to solve this problem using somewhat of a new method. Subsequently James Shearer has been able to solve it using old methods. The answer is 4/13.

The idea of the standard approach is to deduce the minimum intersection of a set of blocking squares with suitably weighted small regions, and deducing a realizable bound from these.

Our method proceeds by associating gaps of unoccupied squares of length three or more with neighboring smaller gaps so as to achieve net density in associated regions of at least 4/13.

Getting back to general operations research questions, there have been many recent efforts in investigating algorithms that have a high probability of success of working or in understanding "average" rather than worst case behavior. And of course there are many efforts to understand the location of the threshold between "hard" and "easy" problems. Given the present state of this subject it seems to me that such questions will not soon be completely resolved, and we can expect many surprising new results in the next few years.

Symplectic Projective Orbits

Bertram Kostant* and Shlomo Sternberg**

Let H be a complex Hilbert space and $P(H)$ the corresponding projective space, i.e., the space of all one dimensional subspaces of H. If v is a non-zero vector in H we shall denote the corresponding point in $P(H)$, i.e., the line through v by $[v]$. Let G be a compact Lie group which is unitarily represented on H so that we may consider the corresponding action of G on $P(H)$. In conjunction with the Hartree–Fock and other approximations, a number of physicists, [2], [4], and [5], have become interested in the following problem: For which smooth vectors v is the orbit $G \cdot [v]$ symplectic? Since G is compact, by projecting onto components we may reduce the problem to the case where the representation is irreducible and hence H is finite dimensional. We restate the question in this case: The unitary structure on H makes $P(H)$ into a Kaehler manifold and thus, in particular, into a symplectic manifold. Each G orbit in $P(H)$ is a submanifold. The question is: for which G orbits is the restriction of the symplectic form of $P(H)$ non-degenerate so that the orbit becomes a symplectic manifold? We shall show that the only symplectic orbits are orbits through projectivized weight vectors. But not all such orbits are symplectic; there is a further restriction on the weight vector that we shall describe. The orbit through the projectivized maximal weight vector is not only symplectic but is also Kaehler, i.e., is a complex submanifold of $P(H)$ and is the only orbit with this property.

For example, let $G = SU(2)$ and $H = \mathbb{C}^5$ with the standard five dimensional (spin 2) irreducible representation. Then $P(H)$ is a four complex dimensional Kaehler manifold and hence an eight (real) dimensional symplectic manifold while $SU(2)$ is a real three dimensional Lie group. A typical orbit in $P(H)$ will be three dimensional and so certainly not symplectic. The

* Department of Mathematics, Massachusetts Institute of Technology, Cambridge, MA 02139.
** Department of Mathematics, Harvard University, Cambridge, MA 02138.

orbits through projectivized weight vectors will be two dimensional spheres and there will be three of them corresponding to the weights $j = 2$, 1, and 0. (The value $-j$ gives the same orbit as j.) The $j = 2$ sphere is a complex submanifold (and so Kaehler and so symplectic). The $j = 1$ sphere is symplectic but not complex. The $j = 0$ sphere is not symplectic (and is in fact Lagrangian—the restriction of the symplectic form vainshes identically on this sphere).

We wish to thank Professors Kramer, Rosensteel, and Rowe for calling this problem to our attention and explaining its physical significance to us.

Now to the details. Let H be a finite dimensional Hilbert space. We may identify $P(H)$ with $S(H)/S^1$, where $S(H)$ is the unit sphere, i.e., is the set of all vectors of length one, and we identify two such vectors as they differ by a phase factor. Each $v \in S(H)$ defines the one dimensional self adjoint projection operator $v \otimes v^*$ which is just projection onto the line through v. Two vectors differing by a phase factor define the same projection. We thus have a map $\psi\colon P(H) \to SA(H)$, where $SA(H)$ denotes the space of all self adjoint operators on H given by

$$\psi([v]) = v \otimes v^*.$$

The space $SA(H)$ has a natural scalar product given by $\frac{1}{2}\operatorname{tr} S_1 S_2$ for $S_i \in SA(H)$. The map ψ pulls the Euclidean metric on $SA(H)$ back to a Riemannian metric on $P(H)$ which we will denote by q. If A and B are linear transformations of H, let $A_{[v]}$ and $B_{[v]}$ denote the corresponding tangent vectors at $[v]$ so that $q_v(A_{[v]}, B_{[v]})$ is the scalar product of these tangent vectors under the Riemannian metric, q. We can evaluate this scalar product as follows:

$$d\psi(A_{[v]}) = Av \otimes v^* + v \otimes A^*v^*$$

so

$$\begin{aligned} q_v(A_{[v]}, B_{[v]}) &= \tfrac{1}{2}\operatorname{tr}(Av \otimes v^* + v \otimes A^*v^*)(Bv \otimes v^* + v \otimes B^*v^*) \\ &= \operatorname{Re}(\langle Av, v\rangle\langle v, Bv\rangle + \langle Av, Bv\rangle) \end{aligned}$$

where $\langle\ ,\ \rangle$ denotes the scalar product on H. It is easy to check that the Riemannian is invariant under the action of the unitary group $U(H)$ of H, and it is clear that it is invariant under multiplication by i in each tangent space, i.e., that $q_v(i\xi, i\eta) = q(\xi, \eta)$. This last condition implies that the bilinear form ω_v defined by

$$\omega_v(\xi, \eta) = q_v(i\xi, \eta)$$

is antisymmetric so that we have defined a two form ω on $P(H)$. One checks that q is positive definite which implies that ω is non-singular, and that $d\omega = 0$, cf., for example [3] pp. 85–87. A Riemannian metric on a complex manifold which is invariant under multiplication by i and which has the property that the associated antisymmetric form is symplectic is called a

Kaehler metric. We have described the unique (up to constant multiple) Kaehler metric on $P(H)$ invariant under $U(H)$.

If A and B are skew adjoint operators, then $\langle iAv, v\rangle$ is real and $\langle v, Bv\rangle$ is imaginary, so the first term in the above expression for $q_v(iA_{[v]}, B_{[v]})$ vanishes and we have

$$\omega_v(A_{[v]}, B_{[v]}) = \operatorname{Im}\langle Av, Bv\rangle = -\frac{i}{2}\langle [A, B]v, v\rangle. \tag{1}$$

Let $u(H)$ denote the Lie algebra of $U(H)$ and $u(H)^*$ its dual space. The second equality in (1) shows that if we define the map Φ: $P(H) \to u(H)^*$ by

$$\Phi([v]) = \frac{i}{2}\langle v, Cv\rangle$$

then

$$\Phi([v])([A, B]) = -\omega_v(A_{[v]}, B_{[v]}).$$

The map Φ is thus a moment map for the action of $U(H)$ on $P(H)$. If G is any Lie group, a unitary representation of G on H means a homomorphism of G into $U(H)$ and this induces a homomorphism, call it v of g into $u(H)$ where g is the Lie algebra of G. We get the dual map $v^* : u(H)^* \to g^*$ and $v^* \circ \Phi$ is a moment map for the induced action of G on $P(H)$. In particular, this action is Hamiltonian and therefore by [1], any symplectic orbit in $P(H)$ must be locally isomorphic to an orbit of G acting on g^*.

If G is compact we may identify g with g^* by an invariant bilinear form. The stabilizer of a point in g^* is the same as the centralizer of a point in g and hence contains some maximal torus. Thus, if $G \cdot [v]$ is symplectic, $[v]$ must be stabilized by some maximal torus T. This means that the line through v is carried into itself by T, i.e., that v is a weight vector relative to T. We have thus proved

Proposition 1. *If $G \cdot [v]$ is a symplectic submanifold of $P(H)$ then v is a weight vector of G relative to some maximal torus, T.*

(Recall that all maximal tori are conjugate and that if $[v]$ is stabilized by T then $a[v]$ is stabilized by aTa^{-1} for any $a \in G$.)

We must now determine which weight vectors give rise to symplectic orbits. Let t denote the Lie algebra of T and $g^{\mathbb{C}}$ the complexification of g. We have a decomposition

$$g^{\mathbb{C}} = t^{\mathbb{C}} + \oplus \mathbb{C} E_\alpha$$

where E_α denotes a root vector corresponding to the root α and α ranges over all the roots. Similarly, we have the decomposition

$$g = t + \oplus \mathbb{R}(E_\alpha - E_{-\alpha}) + \oplus \mathbb{R} i(E_\alpha + E_{-\alpha}).$$

Suppose that v is a weight vector with weight λ. Then $[E_\alpha, E_\beta]v$ is a weight vector of weight $\lambda + \alpha + \beta$ and hence orthogonal to v unless $\alpha = -\beta$. It follows from (1) that the spaces of tangent vectors

$$(E_\alpha - E_{-\alpha})_{[v]}, \qquad i(E_\alpha + E_{-\alpha})_{[v]}$$

are mutually orthogonal with respect to ω_v as α ranges over the set of positive roots. Some of these tangent vectors might be zero. To check whether the orbit is symplectic we need to know that if ω_v vanishes on this subspace then the tangent vectors are zero. Now $[E_\alpha, E_{-\alpha}] = r_\alpha \in t$ and $r_\alpha v = (\lambda \cdot \alpha)v$ where $\lambda \cdot \alpha$ denotes the value of λ on r_α. We have thus proved

Theorem. *An orbit $G \cdot [v]$ in $P(H)$ is symplectic if and only if v is a weight vector satisfying the following condition: If λ is the weight corresponding to v then $\lambda \cdot \alpha = 0$ implies that $E_\alpha v = 0$ for every root α.*

(In particular, regular weights, i.e., those λ for which $\lambda \cdot \alpha \neq 0$ for any α, give rise to symplectic orbits while the zero weight never gives rise to a symplectic orbit (unless the orbit is a point).)

The description of the Kaehler orbits is essentially a consequence of the Borel–Weil theorem. If the orbit were a complex submanifold of $P(H)$, its tangent space would be stable under multiplication by i and so we would get an action of $g^{\mathbb{C}}$ and hence of the complex group $G^{\mathbb{C}}$ on the orbit. The only compact Kaehler homogeneous spaces for $g^{\mathbb{C}}$ are of the form $G^{\mathbb{C}}/P$ where P contains a Borel subgroup. Thus $[v]$ is stabilized by a Borel subgroup and so v is a maximal weight vector. Thus

Proposition 2. *There is only one Kaehler orbit and it is the orbit of a projectivized maximal weight vector.*

References

[1] B. Kostant, Quantization and unitary representations, *Lecture Notes in Math.* **170** (1970), 87–208.

[2] P. Kramer and M. Saraceno, Geometry of the time dependent variational principle in quantum mechanics, to appear.

[3] D. Mumford, *Algebraic Geometry I, Complex Projective Varieties*, Springer-Verlag, Berlin, New York, 1976.

[4] G. Rosensteel, Hartree–Fock–Bogoliubov theory without quasiparticle vacua, to appear.

[5] D. J. Rowe, A. Ryman and G. Rosensteel, Many-body quantum mechanics as a dynamical system, to appear.

Four Applications of Nonlinear Analysis to Physics and Engineering

Jerrold E. Marsden*†

Introduction

My goal is to describe, in as accessible terms as possible, four separate applications of nonlinear analysis to relativity, elasticity, chaotic dynamics and control theory that I have recently been involved with. The descriptions are in some sense superficial since many interesting technical points are glossed over. However, this is necessary to efficiently convey the flavor of the methods.

Most applications of mathematics to "real-life" problems of immediate need do not involve deep methods and ideas. For example, the force exerted on an aircraft frame by the landing gear when the vehicle lands is best computed, at least at first, by using undergraduate mathematics, engineering and experience. However applied mathematics in the broad sense ranges from such problems of urgency to "practical" problems involving deeper mathematics (compute the lift and flutter characteristics for a design modification of the 747) through to fundamental physical problems involving interactions with the frontier of mathematics that need not be of any immediate "need" (is turbulence predictable from the Navier Stokes equations alone?).

The applications I shall speak about are of the fundamental kind involving current research in mathematics and basic questions in physics and engineering that are normally not considered "practical." Most, if not all, of the other lectures I have heard at this conference fall into the same category.

* Department of Mathematics, University of California, Berkeley, CA 94720.

† Research partially supported by NSF grant MCS 78-06718 and ARO grant DAAG 29-79C-0086.

I have heard many arguments about what is and what is not "applied" mathematics, and have seen rifts between individuals and whole departments over this issue. For example, to some, general relativity is not applied mathematics, but quantum mechanics is. To others, even the most abstract continuum mechanics or control theory is applied mathematics while functional analysis or differential geometry used in any subject disqualifies that endeavor from being applied. This would all be very humerous if individuals did not take it so seriously. The results described below are "applied" if the term is used in its broad sense.

Space does not allow for the presentation of an accurate historical picture of each problem, nor for a thorough citation of other approaches. Most of this can, however, be tracked down by consultation of the literature which is cited at the end of the paper.

1. Spaces of Solutions in Relativistic Field Theories*

1.1 Vacuum Gravity

A *spacetime* is a four dimensional manifold V together with a pseudo-Riemannian tensor field g of signature $(+, +, +, -)$. Let $\mathrm{Riem}(g)$ denote the Riemannian–Christoffel curvature tensor computed from g. Relative to a chosen basis in the tangent space to V at a point $x \in V$, $\mathrm{Riem}(g)$ is given in terms of a four-index object denoted $R^{\alpha}{}_{\beta\gamma\delta}$. By contracting two indices, we construct the Ricci curvature $\mathrm{Ric}(g)$ (in coordinates $R_{\alpha\beta}$) and the scalar curvature $R(g)$ (in coordinates $R = R^{\alpha}{}_{\alpha}$. The Einstein tensor is defined by $\mathrm{Ein}(g) = \mathrm{Ric}(g) - \frac{1}{2}R(g)g$ (in coordinates, $G_{\alpha\beta} = R_{\alpha\beta} - \frac{1}{2}Rg_{\alpha\beta}$). The *Einstein equations* for vacuum gravity are simply that $\mathrm{Ein}(g) = 0$ (which is equivalent to $\mathrm{Ric}(g) = 0$).

Let V be fixed and let $\mathscr{E}$ be the set of *all* g's that satisfy the Einstein equations (plus some additional technical smoothness conditions). Let $g_0 \in \mathscr{E}$ be a given solution. We ask: what is the structure of $\mathscr{E}$ in the neighborhood of g_0?

There are two basic reasons why this question is asked. First of all, it is relevant to the problem of finding solutions to the Einstein equations in the form of a perturbation series:

$$g(\lambda) = g_0 + \lambda h_1 + \frac{\lambda^2}{2} h_2 + \cdots$$

where λ is a small parameter. If $g(\lambda)$ is to solve $\mathrm{Ein}(g(\lambda)) = 0$ identically in λ then clearly h_1 must satisfy the *linearized Einstein equations*:

$$D\,\mathrm{Ein}(g) \cdot h_1 = 0$$

* This section is based on joint work with J. Arms, A. Fischer and V. Moncrief.

where D Ein(g) is the derivative of the mapping $g \mapsto \text{Ein}(g)$. For such a perturbation series to be possible, is it sufficient that h_1 satisfy the linearized Einstein equations? i.e., is h_1 necessarily a direction of *linearized stability*? We shall see that in general the answer is no, unless drastic additional conditions hold. The second reason why the structure of $\mathscr{E}$ is of interest is in the problem of quantization of the Einstein equations. Whether one quantizes by means of direct phase space techniques (due to Dirac, Segal, Souriau, and Kostant in various forms) or by Feynman path integrals, there will be difficulties near places where the space of classical solutions is such that the linearized theory is *not* a good approximation to the nonlinear theory.

For vacuum gravity, let us state the answer in a special case: suppose g_0 has a *compact* spacelike hypersurface $M \subset V$. (Technically, M should be a Cauchy surface and be deformable to a surface of constant mean curvature.) Let $\mathscr{S}_{g_0}$ be the Lie group of isometries of g_0 and let k be its dimension.

Theorem

(1) *If $k = 0$, then $\mathscr{E}$ is a smooth manifold in a neighborhood of g_0 with tangent space at g_0 given by the solutions of the linearized Einstein equations.*
(2) *If $k > 0$ then $\mathscr{E}$ is* not *a smooth manifold at g_0. A solution h_1 of the linearized equations is tangent to a curve in $\mathscr{E}$ if and only if h_1 is such that the Taub conserved quantities vanish; i.e., for every Killing field X for g_0,*

$$\int_M X \cdot [D^2 \operatorname{Ein}(g_0) \cdot (h_1, h_1)] \cdot Z \, d\mu_M = 0$$

where Z is the unit normal to the hypersurface M, "$\cdot$" denotes contraction with respect to the metric g_0 and μ_M is the volume element on M.

All explicitly known solutions possess symmetries, so while (1) is "generic," (2) is what occurs in examples. This theorem gives a complete answer to the perturbation question: such a perturbation series is possible if and only if all the Taub quantities vanish.

Let us give a brief abstract indication of why such second order conditions should come in. Suppose X and Y are Banach spaces and $F\colon X \to Y$ is a smooth map. Suppose $F(X_0) = 0$ and $x(\lambda)$ is a curve with $x(0) = x_0$ and $F(x(\lambda)) \equiv 0$. Let $h_1 = x'(0)$ so by the chain rule $DF(x_0) \cdot h_1 = 0$. Now suppose $DF(x_0)$ is not surjective and in fact suppose there is a linear functional $l \in Y^*$ orthogonal to its range: $\langle l, DF(x_0) \cdot u \rangle = 0$ for all $u \in X$. By differentiating $F(x(\lambda)) = 0$ twice at $\lambda = 0$, we get

$$D^2F(x_0) \cdot (h_1, h_1) + DF(x_0) \cdot x''(0) = 0.$$

Applying l gives

$$\langle l, D^2F(x_0) \cdot (h_1, h_1) \rangle = 0$$

which are necessary second order conditions that must be satisfied by h_1.

It is by this general method that one arrives at the Taub conditions. The issue of whether or not these conditions are sufficient is much deeper, requiring extensive analysis and bifurcation theory (for $k = 1$, the Morse lemma is used, while for $k > 1$ the Kuranishi deformation theory is needed).

1.2 General Field Theories

Is the above phenomenon a peculiarity about vacuum gravity or is it part of a more general fact about relativistic field theories? The examples which have been and are being worked out suggest that the latter is the case. Good examples are the Yang–Mills equations for gauge theory, the Einstein–Dirac equations, the Einstein–Euler equations and super-gravity. In such examples there is a gauge group playing the role of the diffeomorphism group of spacetime for vacuum gravity. This gauge group acts on the fields; when it fixes a field, it is a *symmetry* for that field. The relationship between symmetries of a field and singularities in the space of solutions of the classical equations is then as it is for vacuum gravity.

For this program to carry through, one first writes the four dimensional equations as Hamiltonian evolution equations plus constraint equations by means of the $3 + 1$ procedures of Dirac. The constraint equations then must (1) be the Noether conserved quantities for the gauge group and (2) satisfy some technical ellipticity conditions. For (1) it may be necessary to shrink the gauge group somewhat, especially for spacetimes that are not spatially compact. (For example, the isometries of Monkowski space do not belong to the gauge group generating the constraints but rather they generate the total energy-momentum vector of the spacetime.)

1.3 Momentum Maps

The role of the constraint equations as the zero set of the Noether conserved quantity of the gauge group leads one to investigate zero sets of the conserved quantities associated with symmetry groups rather generally. This topic is of interest not only in relativistic field theories, but in classical mechanics too. For example the set of points in the phase space for n particles in $\mathbb{R}^3$ corresponding to zero total angular momentum in an interesting and complicated set, even for $n = 2$!

We shall present just a hint of the relationship between singularities and symmetries. The full story is a long one; one finally ends up with an answer similar to that in relativity.

First we need a bit of notation. Let M be a manifold and let a Lie group G act on M. Associated to each element ξ in the Lie algebra $\mathfrak{g}$ of G, we have a vector field ξ_M naturally induced on M. We shall denote the action by

$\Phi: G \times M \to M$ and we shall write $\Phi_g: M \to M$ for the transformation of M associated with the group element $g \in G$. Thus

$$\xi_M(x) = \frac{d}{dt} \Phi_{\exp(t\xi)}(x) \big|_{t=0}.$$

Now let (P, ω) be a symplectic manifold, so ω is a closed nondegenerate two-form on P and let Φ be an action of a Lie group G on P. Assume the action is symplectic, i.e., $\Phi_g^* \omega = \omega$ for all $g \in G$. A *momentum mapping* is a smooth mapping $J: P \to \mathfrak{g}^*$ such that

$$\langle dJ(x) \cdot v_x, \xi \rangle = \omega_x(\xi_p(x), v_x)$$

for all $\xi \in \mathfrak{g}$, $v_x \in T_x P$ where $dJ(x)$ is the derivative of J at x, regarded as a linear map of $T_x P$ to $\mathfrak{g}^*$ and $\langle , \rangle$ is the natural pairing between $\mathfrak{g}$ and $\mathfrak{g}^*$.

A momentum map is *Ad*-equivariant* when the following diagram commutes for each $g \in G$:

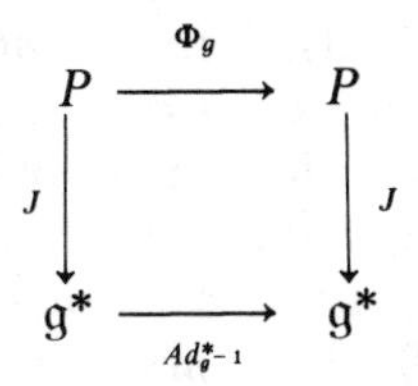

where $Ad^*_{g^{-1}}$ denotes the co-adjoint action of G on $\mathfrak{g}^*$. If J is Ad^* equivariant, we call (P, ω, G, J) a *Hamiltonian G-space*.

Momentum maps represent the (Noether) conserved quantities associated with symmetry groups on phase space. This topic is of course a very old one, but it is only with more recent work of Souriau and Kostant that a deeper understanding has been achieved.

Let $\mathscr{S}_{x_0}$ = (the component of the identity of) $\{g \in G \mid gx_0 = x_0\}$, called the symmetry group of x_0. Its Lie algebra is denoted $\mathfrak{s}_x$, so

$$\mathfrak{s}_{x_0} = \{\xi \in \mathfrak{g} \mid \xi_P(x_0) = 0\}.$$

Let (P, ω, G, J) be a Hamiltonian G-space. If $x_0 \in P$, $\mu_0 = J(x_0)$ and if

$$dJ(x_0)\colon T_x P \to \mathfrak{g}^*$$

is surjective (with split kernel), then locally $J^{-1}(\mu_0)$ is a manifold and $\{J^{-1}(\mu) \mid \mu \in \mathfrak{g}^*\}$ forms a regular local foliation of a neighborhood of x_0. Thus, when $dJ(x_0)$ fails to be surjective, the set of solutions of $J(x) = 0$ could fail to be a manifold.

Theorem. *$dJ(x_0)$ is surjective if and only if* $\dim \mathscr{S}_{x_0} = 0$; *i.e.,* $\mathfrak{s}_{x_0} = \{0\}$.

PROOF. $dJ(x_0)$ fails to be surjective iff there is a $\xi \neq 0$ such that $\langle dJ(x_0) \cdot v_{x_0}, \xi \rangle = 0$ for all $v_{x_0} \in T_{x_0} P$. From the definition of momentum map, this is equivalent to $\omega_{x_0}(\xi_P(x_0), v_{x_0}) = 0$ for all v_{x_0}. Since ω_{x_0} is non-degenerate, this is, in turn, equivalent to $\xi_P(x_0) = 0$; i.e., $\mathfrak{s}_{x_0} \neq \{0\}$. □

One then goes on to study the structure of $J^{-1}(\mu_0)$ when x_0 *has* symmetries, by investigating second order conditions and using methods of bifurcation theory. It turns out that, as in relativistic field theories, $J^{-1}(\mu_0)$ has quadratic singularities characterized by the vanishing of second order conditions. The connection is not an accident since the structure of the space of solutions of a relativistic field theory is determined by the vanishing of the momentum map associated with the gauge group of that theory.

2. The Traction Problem in Nonlinear Elasticity*

2.1 Terminology from Elasticity

Let $\mathscr{B} \subset \mathbb{R}^3$ be an open set with smooth boundary. We regard $\mathscr{B}$ as a reference state for an elastic body. A *configuration* or *deformation* of $\mathscr{B}$ is a (smooth) embedding $\phi: \mathscr{B} \to \mathbb{R}^3$. Let $\mathscr{C}$ denote all such ϕ's. The derivative of ϕ is denoted $F = D\phi$ and is called the *deformation gradient*. The body's elastic properties are characterized by a *stored energy function*, a function W of $X \in \mathscr{B}$ and 3×3 matrices. Thus, given $\phi \in \mathscr{C}$, we get a function of X by the composition $W(X, F(X))$. The (*first*) *Piola–Kirchhoff stress tensor* is defined by $T = \partial W/\partial F$, the derivative with respect to the second argument of W. We shall assume that the undeformed state is stress-free; i.e., $T = 0$ when ϕ = identity.

Let B: $\mathscr{B} \to \mathbb{R}^3$ denote a given *body force* (per unit volume) and $\tau: \partial\mathscr{B} \to \mathbb{R}^3$ a given *surface traction* (per unit area). The equilibrium equations for ϕ we shall study are

$$\begin{aligned} \text{DIV } T + B &= 0 \quad \text{in } \mathscr{B} \\ T \cdot N &= \tau \quad \text{on } \partial\mathscr{B}. \end{aligned} \tag{E}$$

These equations are equivalent to finding the critical points in $\mathscr{C}$ of the energy:

$$V(\phi) = \int_{\mathscr{B}} W\, dV + \int_{\mathscr{B}} \phi \cdot B\, dV + \int_{\partial\mathscr{B}} \phi \cdot \tau\, dA.$$

Let $\mathscr{L}$ be the space of pairs $l = (B, \tau)$ of loads such that

$$\int_{\mathscr{B}} B(X)\, dV(X) + \int_{\partial\mathscr{B}} \tau(X)\, dA(X) = 0,$$

i.e., the total force is zero. By the divergence theorem, if l is a set of loads satisfying the equilibrium equations for some ϕ, then $l \in \mathscr{L}$.

* Based on joint work with D. R. J. Chillingworth and Y. H. Wan.

2.2 Discussion of the Traction Problem

If we were studying the displacement problem i.e., the boundary condition was ϕ prescribed on $\partial\mathscr{B}$, it would follow directly from the implicit function theorem that for any B near zero, there would be a unique ϕ near the identity satisfying the equilibrium equations. For the traction problem the kernel of the linearized equations consists of infinitesimal rigid body motions and the implicit function theorem fails. In fact, the solution set bifurcates near the identity and the geometry of the rotation group $SO(3)$ plays a crucial role. We can trivially remove the translations by specifying the image of a given point in $\mathscr{B}$, say $\phi(0) = 0$.

Our problem is to study the solutions of the equations (E) for various l. The methods by which we do this are those of bifurcation theory and singularity theory. Interestingly, the solutions, even for small l, can be as complex as those for the buckling of a plate with 9 or more nearby solutions.

That there are such difficulties with the traction problem was noticed in the 1930's by Signorini. The problem has been extensively studied by the Italian school, especially by Stoppelli. However, their analysis missed solutions because the methods used are not "robust;" i.e., they did not allow the loads to move in full neighborhoods (they did not include enough parameters). Moreover, others were missed because the global geometry of $SO(3)$ was not exploited. Finally, the stability of the various solutions was not obtained.

We shall give just a hint of our methods by sketching a new and much simplified proof of a theorem of Stoppelli in case there is no bifurcation.

Let $\Phi: \mathscr{C} \to \mathscr{L}$ be defined by

$$\Phi(\phi) = (-\text{DIV } T, T \cdot N)$$

so the equilibrium equations are $\Phi(\phi) = l$.

Let

$$\mathscr{C}_l = \{u \in T_{id}\mathscr{C} \mid u(0) = 0 \quad \text{and} \quad Du(0) \text{ is symmetric}\}$$

and let the *equilibrated loads* be those whose torque in the reference configuration is zero, i.e.,

$$\mathscr{L}_e = \left\{ l \in \mathscr{L} \,\middle|\, \int_\Omega X \times B(X)\, dV(X) + \int_{\partial\Omega} X \times \tau(X)\, dA(X) = 0 \right\}.$$

Assuming the appropriate ellipticity conditions from linear elasticity, we know that

$$D\Phi(id)|_{\mathscr{C}_l}: \mathscr{C}_l \to \mathscr{L}_e$$

is an isomorphism.

Let $SO(3)$ act on $\mathscr{C}$ and $\mathscr{L}$ in the obvious way: For $Q \in SO(3)$, $\phi \in \mathscr{C}$ and $l \in \mathscr{L}$, let

$$(Q, \phi) \mapsto Q \circ \phi \qquad \text{and} \qquad (Q, l) \mapsto (Q \circ \beta, Q \circ \tau).$$

For $l \in \mathscr{L}$, let $\mathcal{O}_l$ denote the $SO(3)$ orbit of l:

$$\mathcal{O}_l = \{Ql \mid Q \in SO(3)\}.$$

Let $l \in \mathscr{L}_e$. Then l is said to have *no axis of equilibrium* if, for all $\xi \in SO(3)$, $\xi \neq 0$ we have

$$\xi l \notin \mathscr{L}_e,$$

i.e., any rotation of l destroys the equilibration. If l has an axis of equilibrium, then there is a vector $e \in \mathbb{R}^3$ such that rotations of l about e map l into $\mathscr{L}_e$, as is readily checked.

Lemma (Da Silva's Theorem). *Let* $l \in \mathscr{L}$. *Then* $\mathcal{O}_l \cap \mathscr{L}_e \neq \varnothing$.

PROOF. Define the *astatic load map* $k: \mathscr{L} \to M_3$(3 × 3 matrices) by

$$k(l) = K(B, \tau) = \int_\Omega X \otimes B(X)\, dV(X) + \int_{\partial\Omega} X \otimes \tau(X)\, dA(X)$$

so that $l \in \mathscr{L}_e$ iff $k(l)$ is symmetric. Now k is $SO(3)$ equivariant:

$$\begin{array}{ccc} \mathscr{L} & \xrightarrow{\;k\;} & M_3 \\ {\scriptstyle SO(3)}\downarrow & & \downarrow{\scriptstyle SO(3)} \\ \mathscr{L} & \xrightarrow{\;k\;} & M_3 \end{array}$$

where the action on M_3 is $(Q, A) \mapsto AQ^{-1}$, i.e.,

$$k(Ql) = k(l)Q^{-1}.$$

The result is now obvious from the polar decomposition. □

We also assume that Φ is equivariant (called *material frame indifference*):

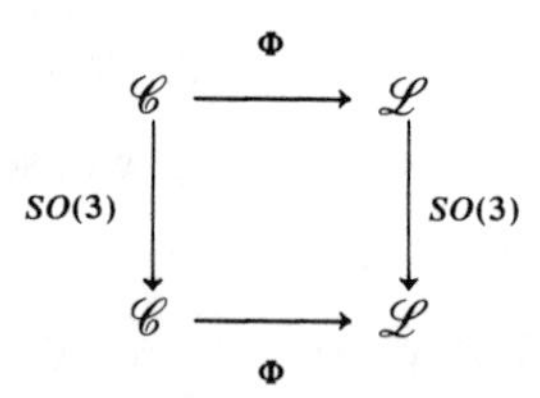

Thus, to study the solutions of $\Phi(\phi) = l$ for a given l, we can assume that $l \in \mathscr{L}_e$.

2.3 A Proof of Existence and Uniqueness in the Simplest Case

Suppose now that $l \in \mathscr{L}_e$ is given and has no axis of equilibrium. The main theorem in this case is due to Stoppelli which we now prove.

Lemma (a) $\dim \mathcal{O}_l = 3$ *and* (b) $T_l\mathcal{O}_l \oplus \mathscr{L}_e = \mathscr{L}$.

PROOF. If dim $\mathcal{O}_l < 3$, there would be a $\xi \neq 0$, $\xi \in SO(3)$ such that $\xi l = 0$, which contradicts $\xi l \notin \mathcal{L}_e$. Thus (a) holds. Also, by the no axis of equilibrium assumption, $T_l \mathcal{O}_l \cap \mathcal{L}_e = \{0\}$. Since $\mathcal{L}_e$ has codimension 3 in $\mathcal{L}$ and (a) holds, we get (b). □

Let $\tilde{\Phi}$ be the restriction of Φ to $\mathcal{C}_l$, regarded as an affine subspace of $\mathcal{C}$ centered at the identity. As remarked before,

$$D\tilde{\Phi}(id): \mathcal{C}_l \to \mathcal{L}_e$$

is an isomorphism. In particular, it is one to one and so for ϕ in a neighborhood of the identity

$$\text{Range } \tilde{\Phi} \equiv N$$

is a submanifold of $\mathcal{L}$ tangent to $\mathcal{L}_e$ at the origin (see Figure 1). By the above lemma,

$$\{Ql \mid Q \in \text{a neighborhood } U \text{ of } Id \in SO(3)\}$$

is a neighborhood of l in the normal direction to $\mathcal{L}_e$. Thus

$$\{\lambda Ql \mid Q \in U, \lambda \in (-\varepsilon, \varepsilon)\}$$

is the cone in the normal bundle to $\mathcal{L}_e$.

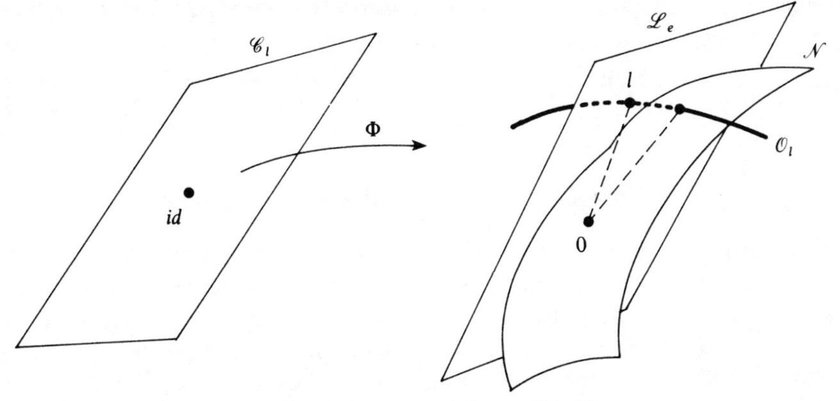

Figure 1 The Geometry of Stoppelli's Theorem

Since N is tangent to $\mathcal{L}_e$ at 0, for λ sufficiently small $\mathcal{O}_{\lambda l}$ will intersect N. Thus, for λ sufficiently small, there is a unique Q in a neighborhood of the Identity such that

$$\Phi(\bar{\phi}) = \lambda Ql$$

has a unique solution $\bar{\phi} \in \mathcal{C}_l$. Thus $\phi = Q^{-1}\bar{\phi}$ solves $\Phi(\phi) = \lambda l$. Thus we have proved:

Theorem (Stoppelli). *Suppose $l \in \mathcal{L}_e$ has no axis of equilibrium. Then for λ*

sufficiently small, there is a unique $\bar{\phi} \in \mathscr{C}_l$ *and* Q *in a neighborhood of the identity such that* $\phi = Q^{-1}\bar{\phi}$ *solves the traction problem:*

$$\Phi(\phi) = \lambda l.$$

2.4 Discussion of the General Case

The main problem is to study the situation when l is near a load l_0 with an axis of equilibrium. To do so one must first classify how degenerate the axis of equilibrium is. This is done by classifying how the orbits of the action of $SO(3)$ on M_3 meet Sym, the symmetric matrices. There are five such types. For example, if $A \in M_3$ has no axis of equilibrium and has distinct eigenvalues, then $\mathcal{O}_A$ meets Sym transversally in four points (Type 0). If A, however, has no axis of equilibrium and two equal non-zero eigenvalues, $\mathcal{O}_A$ meets Sym transversally in two points (with no axis of equilibrium) and a circle each point of which has an axis of equilibrium (Type 1). If A has a triple non-zero eigenvalue, $\mathcal{O}_A$ meets Sym transversally in one point (A itself) and in an $\mathbb{RP}^2$, each point of which has a circle of axes of equilibrium (Type 2). There are also the more degenerate types 3 and 4.

When the Liapunov–Schmidt procedure from bifurcation theory is applied to this situation, one ends up with a bifurcation problem of vector fields on S^1 for type 1 and of vector fields on $\mathbb{RP}^2$ for type 2. These can then be analyzed by singularity theory and one finds cusps and double cusps respectively. Previously, the best that was known was due to Stoppelli: he saw only particular sections of the cusps in type 1 and did not analyze type 2.

3. Chaotic Oscillations of a Forced Beam*

The study of chaotic motion in dynamical systems is now a burgeoning industry. The literature is currently in a state of explosion. We shall sketch an example from structural mechanics for which one can prove that the associated dynamical system has complex dynamics. Part of the interest is that methods of ordinary differential equations can be made to work for a certain class of partial differential equations.

We shall state the result for the main example first and then sketch the abstract theory which is used for the proof.

3.1 The Main Example

Consider a beam that is buckled by an external load Γ, so that there are two stable and one unstable equilibrium states (see Figure 2). The whole structure is shaken with a transverse periodic displacement, $f \cos \omega t$, and the

* Based on joint work with P. Holmes

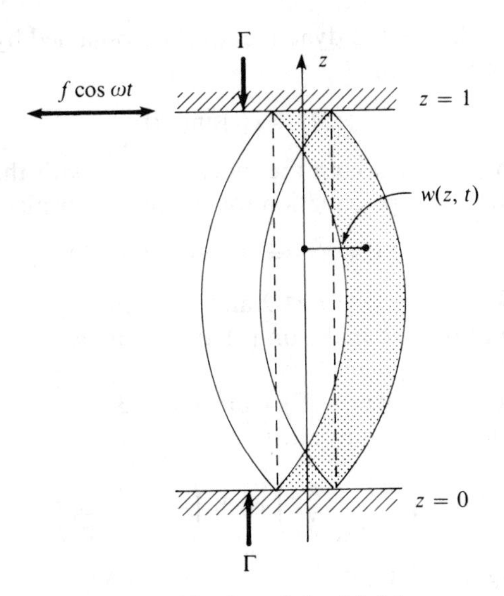

Figure 2 The forced, buckled beam

beam moves due to its inertia. One observes periodic motion about either of the two stable equilibria for small f, but as f is increased, the motion becomes aperiodic or chaotic.

A specific model for the transverse deflection $w(z, t)$ of the centerline of the beam is the following partial differential equation:

$$\ddot{w} + w'''' + \Gamma w'' - \kappa\left(\int_0^1 [w']^2 \, d\zeta\right) w'' = \varepsilon(f \cos \omega t - \delta \dot{w}) \tag{1}$$

where $\cdot = \partial/\partial t$, $' = \partial/\partial z$, Γ = external load, κ = stiffness due to "membrane" effects, δ = damping, and ε is a parameter used to measure the size of f and δ. Amongst many possible boundary conditions we shall choose $w = w'' = 0$ at $z = 0, 1$, i.e., simply supported, or hinged ends. With these boundary conditions, the eigenvalues of the linearized, unforced equations, i.e., complex numbers λ such that

$$\lambda^2 w + w'''' + \Gamma w'' = 0$$

for some non-zero w satisfying $w = w'' = 0$ at $z = 0, 1$, form a countable set

$$\lambda_j = \pm \pi j \sqrt{\Gamma - \pi^2 j^2}, \qquad j = 1, 2, \ldots .$$

Assume that

$$\pi^2 < \Gamma < 4\pi^2,$$

in which case the solution $w = 0$ is unstable with one positive and one negative eigenvalue and the nonlinear equation (1) with $\varepsilon = 0$, $\kappa > 0$ has two nontrivial stable buckled equilibrium states.

A simplified model for the dynamics of (1) is obtained by seeking lowest mode solutions of the form

$$w(z, t) = x(t)\sin(\pi z).$$

Substitution into (1) and taking the inner product with the basis function $\sin(\pi z)$ gives a Duffing type equation for the modal displacement $x(t)$:

$$\ddot{x} - \beta x + \alpha x^3 = \varepsilon(\gamma \cos \omega t - \delta \dot{x}), \tag{2}$$

where $\beta = \pi^2(\Gamma - \pi^2) > 0$, $\alpha = \kappa\pi^4/2$ and $\gamma = 4f/\pi$.

Further assumptions we make on (1) are as follows:

(1) (No resonance): $j^2\pi^2(j^2\pi^2 - \Gamma) \neq \omega^2, j = 2, 3, 4, \ldots$.
(2) (Large forcing to damping ratio):

$$\frac{f}{\delta} > \left(\frac{\pi}{3}\frac{\Gamma - \pi^2}{\omega\sqrt{k}}\right) \cosh\left(\frac{\omega}{2\sqrt{\Gamma - \pi^2}}\right).$$

(3) (Small forcing and damping): ε is sufficiently small.

On an appropriate function space X, one shows that (1) has well-defined dynamics; elements of X are certain pairs $(w, \dot{w})$. In particular, there is a time $2\pi/\omega$ map $P: X \to X$ that takes initial data and advances it in time by one period of the forcing function.

Theorem. *Under the above hypotheses, there is some power P^N of P that has an invariant set $\Lambda \subset X$ on which P^N is conjugate to a shift on two symbols. In particular,* (1) *has infinitely many periodic orbits with arbitrarily high period.*

This set Λ arises in a way similar to Smale's famous "horseshoe."

3.2 Abstract Hypotheses

We consider an evolution equation in a Banach space X of the form

$$\dot{x} = f_0(x) + \varepsilon f_1(x, t) \tag{3}$$

where f_1 is periodic of period T in t. Our hypotheses on (3) are as follows.

(H1) (a) *Assume $f_0(x) = Ax + B(x)$ where A is an (unbounded) linear operator which generates a C^0 one parameter group of transformations on X and where $B: X \to X$ is C^∞. Assume that $B(0) = 0$ and $DB(0) = 0$.*

(*b*) *Assume $f_1: X \times S^1 \to X$ is C^∞ where $S^1 = \mathbb{R}/(T)$, the circle of length T.*

Assumption 1 implies that the associated suspended autonomous system on $X \times S^1$,

$$\begin{aligned} \dot{x} &= f_0(x) + \varepsilon f_1(x, \theta) \\ \dot{\theta} &= 1, \end{aligned} \tag{4}$$

has a smooth local flow, F_t^ε. This means that $F_t^\varepsilon: X \times S^1 \to X \times S^1$ is a smooth map defined for small $|t|$ which is jointly continuous in all variables $\varepsilon, t, x \in X, \theta \in S^1$ and for x_0 in the domain of A, $t \mapsto F_t^\varepsilon(x_0, \theta_0)$ is the unique solution of (4) with initial condition x_0, θ_0.

The final part of assumption 1 follows:

(c) *Assume that F_t^ε is defined for all $t \in \mathbb{R}$ for $\varepsilon > 0$ sufficiently small.*

Our second assumption is that the unperturbed system is Hamiltonian. This means that X carries a skew symmetric continuous bilinear map Ω: $X \times X \to \mathbb{R}$ which is weakly non-degenerate (i.e., $\Omega(u, v) = 0$ for all v implies $u = 0$) called the *symplectic form* and there is a smooth function $H_0: X \to \mathbb{R}$ such that

$$\Omega(f_0(x), u) = dH_0(x) \cdot u$$

for all x in D_A, the domain of A.

(H2) (a) *Assume that the unperturbed system $\dot{x} = f_0(x)$ is Hamiltonian with energy $H_0: X \to \mathbb{R}$.*

(b) *Assume there is a symplectic 2-manifold $\Sigma \subset X$ invariant under the flow F_t^0 and that on Σ the fixed point $p_0 = 0$ has a homoclinic orbit $x_0(t)$, i.e.,*

$$\dot{x}_0(t) = f_0(x_0(t))$$

and

$$\operatorname*{limit}_{t \to +\infty} x_0(t) = \operatorname*{limit}_{t \to -\infty} x_0(t) = 0.$$

Next we introduce a non-resonance hypothesis.

(H3) (a) *Assume that the forcing term $f_1(x, t)$ in (3) has the form*

$$f_1(x, t) = A_1 x + f(t) + g(x, t) \tag{5}$$

where $A_1: X \to X$ is a bounded linear operator, f is periodic with period T, $g(x, t)$ is t-periodic with period T and satisfies $g(0, t) = 0$, $D_x g(0, t) = 0$, so g admits the estimate

$$\|g(x, t)\| \leq (\text{Const})\|x\|^2 \tag{6}$$

for x in a neighborhood of 0.

(b) *Suppose that the "linearized" system*

$$\dot{x}_L = Ax_L + \varepsilon A_1 x_L + \varepsilon f(t) \tag{7}$$

has a T-periodic solution $x_L(t, \varepsilon)$ such that $x_L(t, \varepsilon) = O(\varepsilon)$.

For finite dimensional systems, (H3) can be replaced by the assumption that 1 does not lie in the spectrum of e^{TA}; i.e., none of the eigenvalues of A resonate with the forcing frequency.

Next, we need an assumption that A_1 contributes positive damping and that $p_0 = 0$ is a saddle.

(H4) (a) *For* $\varepsilon = 0$, e^{TA} *has a spectrum consisting of two simple real eigenvalues* $e^{\pm\lambda T}$, $\lambda \neq 0$, *with the rest of the spectrum on the unit circle.*

(b) *For* $\varepsilon > 0$, $e^{T(A+\varepsilon A_1)}$ *has a spectrum consisting of two simple real eigenvalues* $e^{T\lambda_\varepsilon \pm}$ *(varying continuously in* ε *from perturbation theory of spectra) with the rest of the spectrum,* σ_R^ε, *inside the unit circle* $|z| = 1$ *and obeying the estimates*

$$C_2 \varepsilon \leq \text{distance } (\sigma_R^\varepsilon, |z| = 1) \leq C_1 \varepsilon \tag{8}$$

for C_1, C_2 *positive constants.*

Finally, we need an extra hypothesis on the nonlinear term. We have already assumed B vanishes at least quadratically as does g. Now we assume B vanishes cubically.

(H5) $B(0) = 0$, $DB(0) = 0$, *and* $D^2B(0) = 0$.

This means that in a neighborhood of 0,

$$\|B(x)\| \leq \text{Const } \|x\|^3$$

(actually, $B(x) = o(\|x\|^2)$ would do).

3.3 Some Technical Lemmas

Consider the suspended system (4) with its flow $F_t^\varepsilon: X \times S^1 \to X \times S^1$. Let $P^\varepsilon: X \to X$ be defined by

$$P^\varepsilon(x) = \pi_1 \cdot (F_T^\varepsilon(x, 0))$$

where $\pi_1: X \times S^1 \to X$ is the projection onto the first factor. The map P^ε is just the Poincaré map for the flow F_t^ε. Note that $P^0(p_0) = p_0$, and that fixed points of P^ε correspond to periodic orbits of F_t^ε.

Lemma 1. *For* $\varepsilon > 0$ *small, there is a unique fixed point* p_ε *of* P^ε *near* $p_0 = 0$; *moreover* $p_\varepsilon - p_0 = O(\varepsilon)$, *i.e., there is a constant K such that* $\|p_\varepsilon\| \leq K\varepsilon$ *(for all (small)* ε*).*

For ordinary differential equations, Lemma 1 is a standard fact about persistence of fixed points, assuming 1 does not lie in the spectrum of e^{TA} (i.e., p_0 is hyperbolic). For general partial differential equations, the proof is similar in spirit but is more delicate, requiring our assumptions. An analysis of the spectrum yields the following.

Lemma 2. *For* $\varepsilon > 0$ *sufficiently small, the spectrum of* $DP^\varepsilon(p_\varepsilon)$ *lies strictly inside the unit circle with the exception of the single real eigenvalue* $e^{T\lambda_\varepsilon +} > 1$.

The next lemma deals with invariant manifolds.

Lemma 3. *Corresponding to the eigenvalues $e^{T\lambda_\varepsilon^\pm}$ there are unique invariant manifolds $W^{ss}(p_\varepsilon)$ (the strong stable manifold) and $W^u(p_\varepsilon)$ (the unstable manifold) of p_ε for the map p_ε such that*

(i) *$W^{ss}(p_\varepsilon)$ and $W^u(p_\varepsilon)$ are tangent to the eigenspaces of $e^{T\lambda_\varepsilon^\pm}$, respectively, at p_ε;*
(ii) *they are invariant under P^ε;*
(iii) *if $x \in W^{ss}(p_\varepsilon)$ then*

$$\underset{n\to\infty}{\text{limit}}(P^\varepsilon)^n(x) = p_\varepsilon$$

and if $x \in W^u(p_\varepsilon)$ then

$$\underset{n\to-\infty}{\text{limit}}(P^\varepsilon)^n(n) = p_\varepsilon;$$

(iv) *for any finite t^*, $W^{ss}(p_\varepsilon)$ is C^r close as $\varepsilon \to 0$ to the homoclinic orbit $x_0(t)$, $t^* \le t < \infty$ and for any finite t_*, $W^u(p_\varepsilon)$ is C^r close to $x_0(t)$, $-\infty < t \le t_*$ as $\varepsilon \to 0$ (here, r is any fixed integer, $0 \le r < \infty$).*

The Poincaré map P^ε was associated to the section $X \times \{0\}$ in $X \times S^1$. Equally well, we can take the section $X \times \{t_0\}$ to get Poincaré maps $P^\varepsilon_{t_0}$. By definition,

$$P^\varepsilon_{t_0}(x) = \pi_1(F^\varepsilon_T(x, t_0)).$$

There is an analogue of Lemmas 1, 2, and 3 for $P^\varepsilon_{t_0}$. Let $p_\varepsilon(t_0)$ denote its unique fixed point and $W^{ss}_\varepsilon(p_\varepsilon(t_0))$ and $W^u_\varepsilon(p_\varepsilon(t_0))$ be its strong stable and unstable manifolds. Lemma 2 implies that the stable manifold $W^s(p_\varepsilon)$ of p_ε has codimension 1 in X. The same is then true of $W^s(p_\varepsilon(t_0))$ as well.

Let $\gamma_\varepsilon(t)$ denote the periodic orbit of the (suspended) system (4) with $\gamma_\varepsilon(0) = (p_\varepsilon, 0)$. We have

$$\gamma_\varepsilon(t) = (p_\varepsilon(t), t).$$

The invariant manifolds for the periodic orbit γ_ε are denoted $W^{ss}_\varepsilon(\gamma_\varepsilon)$, $W^s_\varepsilon(\gamma_\varepsilon)$ and $W^u_\varepsilon(\gamma_\varepsilon)$. We have

$$W^s_\varepsilon(p_\varepsilon(t_0)) = W^s_\varepsilon(\gamma_\varepsilon) \cap (X \times \{t_0\})$$
$$W^{ss}_\varepsilon(p_\varepsilon(t_0)) = W^{ss}_\varepsilon(\gamma_\varepsilon) \cap (X \times \{t_0\})$$

and

$$W^u_\varepsilon(p_\varepsilon(t_0)) = W^u_\varepsilon(\gamma_\varepsilon) \cap (X \times \{t_0\}).$$

We wish to study the structure of $W^u_\varepsilon(p_\varepsilon(t_0))$ and $W^s_\varepsilon(p_\varepsilon(t_0))$ and their intersections. To do this, we first study the perturbation of solution curves in $W^{ss}_\varepsilon(\gamma_\varepsilon)$, $W^s_\varepsilon(\gamma_\varepsilon)$ and $W^u_\varepsilon(\gamma_\varepsilon)$.

Choose a point, say $x_0(0)$, on the homoclinic orbit for the unperturbed system. Choose a codimension 1 hyperplane H transverse to the homoclinic orbit at $x_0(0)$. Since $W^{ss}_\varepsilon(p_\varepsilon(t_0))$ is C^r close to $x_0(0)$, it intersects H in a unique point, say $x_\varepsilon^s(t_0, t_0)$. Define $(x_\varepsilon^s(t, t_0), t)$ to be the unique integral curve of the suspended system (4) with initial condition $x_\varepsilon^s(t_0, t_0)$. Define $x_\varepsilon^u(t, t_0)$ in a similar way. We have

$$x_\varepsilon^s(t_0, t_0) = x_0(0) + \varepsilon v^s + O(\varepsilon^2)$$

and

$$x_\varepsilon^u(t_0, t_0) = x_0(0) + \varepsilon v^u + O(\varepsilon^2)$$

by construction, where $\|0(\varepsilon^2)\| \leq \text{Constant} \cdot \varepsilon^2$ and v^s and v^u are fixed vectors. Notice that

$$(P_{t_0}^\varepsilon)^n x_\varepsilon^s(t_0, t_0) = x_\varepsilon^s(t_0 + nT, t_0) \to p_\varepsilon(t_0) \quad \text{as} \quad n \to \infty.$$

Since $x_\varepsilon^s(t, t_0)$ is an integral curve of a perturbation, we can write

$$x_\varepsilon^s(t, t_0) = x_0(t - t_0) + \varepsilon x_1^s(t, t_0) + O(\varepsilon^2),$$

where $x_1^s(t, t_0)$ is the solution of the first variation equation

$$\frac{d}{dt} x_1^s(t, t_0) = Df_0(x_0(t - t_0)) \cdot x_1^s(t, t_0) + f_1(x_0(t - t_0), t), \tag{9}$$

with $x_1^s(t_0, t_0) = v^s$.

3.4 The Melnikov Function

Define the *Melnikov function* by

$$\Delta_\varepsilon(t, t_0) = \Omega(f_0(x_0(t - t_0)), x_\varepsilon^s(t, t_0) - x_\varepsilon^u(t, t_0))$$

and set

$$\Delta_\varepsilon(t_0) = \Delta_\varepsilon(t_0, t_0).$$

Lemma 4. *If ε is sufficiently small and $\Delta_\varepsilon(t_0)$ has a simple zero at some t_0 and maxima and minima that are at least $O(\varepsilon)$, then $W_\varepsilon^u(p_\varepsilon(t_0))$ and $W_\varepsilon^s(p_\varepsilon(t_0))$ intersect transversally near $x_0(0)$.*

The idea is that if $\Delta_\varepsilon(t_0)$ changes sign, then $x_\varepsilon^s(t_0, t_0) - x_\varepsilon^u(t_0, t_0)$ changes orientation relative to $f_0(x_0(0))$. Indeed, this is what symplectic forms measure. If this is the case, then as t_0 increases, $x_\varepsilon^s(t_0, t_0)$ and $x_\varepsilon^u(t_0, t_0)$ "cross," producing the transversal intersection.

The next lemma gives a remarkable formula that enables one to explicitly compute the leading order terms in $\Delta_\varepsilon(t_0)$ in examples.

Lemma 5. *The following formula holds:*

$$\Delta_\varepsilon(t_0) = -\varepsilon \int_{-\infty}^{\infty} \Omega(f_0(x_0(t - t_0)), f_1(x_0(t - t_0), t))\, dt + O(\varepsilon^2).$$

PROOF. Write $\Delta_\varepsilon(t, t_0) = \Delta_\varepsilon^+(t, t_0) - \Delta_\varepsilon^-(t, t_0) + O(\varepsilon^2)$, where

$$\Delta_\varepsilon^+(t, t_0) = \Omega(f_0(x_0(t - t_0)), \varepsilon x_1^s(t, t_0))$$

and

$$\Delta_\varepsilon^-(t, t_0) = \Omega(f_0(x_0(t - t_0)), \varepsilon x_1^u(t, t_0)).$$

Using (9), we get

$$\frac{d}{dt}\Delta_\varepsilon^+(t, t_0) = \Omega(Df_0(x_0(t, t_0)) \cdot f_0(x_0(t - t_0)), \varepsilon x_1^s(t, t_0))$$
$$+ \Omega(f_0(x_0(t - t_0)), \varepsilon\{Df_0(x_0(t - t_0)) \cdot x_1^s(t, t_0) + f_1(x_0(t - t_0), t)\}).$$

Since f_0 is Hamiltonian, Df_0 is Ω-skew. Therefore the terms involving x_1^s drop out, leaving

$$\frac{d}{dt}\Delta_\varepsilon^+(t, t_0) = \Omega(f_0(x_0(t - t_0)), \varepsilon f_1(x_0(t - t_0), t)).$$

Integrating, we have

$$-\Delta_\varepsilon^+(t_0, t_0) = \varepsilon \int_{t_0}^{\infty} \Omega(f_0(x_0(t - t_0)), f_1(x_0(t - t_0), t))\, dt,$$

since

$$\Delta_\varepsilon^+(\infty, t_0) = \Omega(f_0(p_0), \varepsilon f_1(p_0, \infty)) = 0, \quad \text{because } f_0(p_0) = 0.$$

Similarly, we obtain

$$\Delta_\varepsilon^-(t_0, t_0) = \varepsilon \int_{-\infty}^{t_0} \Omega(f_0(x_0(t - t_0)), f_1(x_0(t - t_0), t))\, dt$$

and adding gives the stated formula. □

We summarize the situation as follows.

Theorem. *Let hypotheses* (H1)–(H5) *hold. Let*

$$M(t_0) = \int_{-\infty}^{\infty} \Omega(f_0(x_0(t - t_0)), f_1(x_0(t - t_0), t))\, dt.$$

Suppose that $M(t_0)$ has a simple zero as a function of t_0. Then for $\varepsilon > 0$ sufficiently small, the stable manifold $W_\varepsilon^s(p_\varepsilon(t_0))$ of p_ε for $P_{t_0}^\varepsilon$ and the unstable manifold $W_\varepsilon^u(p_\varepsilon(t_0))$ intersect transversally.

Having established the transversal intersection of the stable and unstable manifolds, one can now plug into known results in dynamical systems (going back to Poincaré) to deduce that the dynamics must indeed be complex. In particular, the previous theorem concerning equation (1) may be deduced.

4. A Control Problem for a Beam*

We wish to point out some unexpected peculiarities in a seemingly straight forward control problem. In particular, the naive methods used for ordinary differential equations do not work for the partial differential equation we discuss. The difficulty has to do with controlling all the modes at once. If the energy norm is used, controllability is impossible. However, if a different asymptotic condition on the modes is used, control is possible.

4.1 The General Scheme for Controllability

Things will run smoothest if we treat the abstract situation first. We consider an evolution equation of the form

$$\dot{u}(t) = \mathscr{A}u(t) + p(t)\mathscr{B}(u(t)) \tag{1}$$

where $\mathscr{A}$ generates a C^0 semigroup on a Banach space X, $p(t)$ is a real value function of t that is locally L^1, and $\mathscr{B}: X \to X$ is C^k, $k \geq 1$. The control question we ask is: let u_0 be given initial data for u and let $T > 0$ be given; does there exist a neighborhood U of $e^{\mathscr{A}T}u_0$ in X such that for any $v \in V$ there exists a p such that the solution of (1) with initial data u_0 reaches v after time T? If the answer is yes, we say (1) is *locally controllable* around the free solution $e^{\mathscr{A}t}u_0$.

The obvious way to tackle this problem is to use the implicit function theorem. Write (1) in integrated form:

$$u(t) = e^{\mathscr{A}t}u_0 + \int_0^t e^{(t-s)\mathscr{A}}p(x)\mathscr{B}(u(s))\,ds. \tag{2}$$

Let p belong to a specified Banach space $Z \subset L^1([0, T], \mathbb{R})$. Standard techniques using the contraction mapping theorem show that for short time, (2) has a unique solution $u(t, p, u_0)$ that is C^k in p and u_0. If we assume $\|\mathscr{B}(x)\| \leq C + K\|x\|$ (for example, $\mathscr{B}$ linear will be of interest to us), then solutions are globally defined, so we do not need to worry about taking short time intervals. The choice $p = 0$ corresponds to the free solution $e^{t\mathscr{A}}u_0$. The derivative $L: Z \to X$ of $u(T, p, u_0)$ with respect to p at $p = 0$ is found by implicitly differentiating (2). One gets

$$Lp = \int_0^T e^{(t-s)\mathscr{A}}p(s)\mathscr{B}(e^{s\mathscr{A}}u_0)\,ds. \tag{3}$$

The implicit function theorem then gives:

Theorem. *If $L: Z \to X$ is a surjective linear map, then (1) is locally controllable around the free solution.*

* Based on joint work with J. Ball and M. Slemrod.

For example, if $X = \mathbb{R}^n$ and $\mathscr{B}$ is linear, we can expand

$$e^{-s\mathscr{A}}\mathscr{B}e^{s\mathscr{A}} = \mathscr{B} + s[\mathscr{A}, \mathscr{B}] + \frac{s^2}{2}[\mathscr{A}, [\mathscr{A}, \mathscr{B}]] + \cdots$$

to recover the standard controllability criterion:

$$\dim \operatorname{span}\{\mathscr{B}u_0, [\mathscr{A}, \mathscr{B}]u_0, [\mathscr{A}, [\mathscr{A}, \mathscr{B}]]u_0, \ldots\} = n.$$

If one wishes to only observe a finite dimensional piece of u, the above method is effective in examples. (By this we mean to control Gu, where $G: X \to \mathbb{R}^n$ is a surjective linear map ... this means we control n "modes" of u.) However, even in the simplest examples, L may have dense range but not be onto. We give such an example below.

4.2 Hyperbolic Systems

Let A be a positive self-adjoint operator on a real Hilbert space H with inner product $\langle\ ,\ \rangle_H$. Let A have a spectrum consisting of eigenvalues λ_n^2, $0 < \lambda_1 \leq \lambda_2 \leq \lambda_3 \leq \cdots$ with corresponding orthonormalized eigenfunctions ϕ_n. Let $B: D(A^{1/2}) \to H$ be bounded. We consider the equation

$$\ddot{w} + Aw + pBw = 0.$$

This is in the form (1) with

$$u = \begin{pmatrix} w \\ \dot{w} \end{pmatrix}$$

and

$$\mathscr{A} = \begin{pmatrix} 0 & I \\ A & 0 \end{pmatrix}, \qquad \mathscr{B} = \begin{pmatrix} 0 & 0 \\ -B & 0 \end{pmatrix}.$$

Here $X = D(A^{1/2}) \times H$ and $\mathscr{A}$ generates a C^0 group of isometries on X. The inner product on X is given by the "energy inner product:"

$$\langle (y_1, z_1), (y_2, z_2) \rangle_X = \langle A^{1/2}y_1, A^{1/2}y_2 \rangle_H + \langle z_1, z_2 \rangle_H.$$

Write

$$u_0 = \begin{pmatrix} \sum_{m=1}^{\infty} b_m \phi_m \\ \sum_{m=1}^{\infty} -\lambda_m c_m \phi_m \end{pmatrix} \in X$$

where

$$\sum_{m=1}^{\infty} \lambda_m^2 (b_m^2 + c_m^2) < \infty.$$

If we set $a_m = \frac{1}{2}(b_m + ic_m)$ we have

$$e^{\mathscr{A}s}u_0 = \begin{pmatrix} \sum_{m=1}^{\infty} [a_m \exp(i\lambda_m s) + \bar{a}_m \exp(-i\lambda_m s)]\phi_m \\ \sum_{m=1}^{\infty} i\lambda_m[a_m \exp(i\lambda_m s) - \bar{a}_m \exp(-i\lambda_m s)]\phi_m \end{pmatrix}$$

and

$$\mathscr{B}e^{\mathscr{A}s}u_0 = \begin{pmatrix} 0 \\ \sum_{m=1}^{\infty} [a_m \exp(i\lambda_m s) + \bar{a}_m \exp(-i\lambda_m s)]B\phi_m \end{pmatrix}.$$

To simplify matters, let us assume that $\langle B\phi_m, \phi_n\rangle_H = d_m \delta_{mm}$. Then

$$e^{-s\mathscr{A}}\mathscr{B}e^{s\mathscr{A}}u_0 = \begin{pmatrix} \sum_{n=1}^{\infty} \frac{-id_n}{2\lambda_n} \{a_n \exp(2i\lambda_n s) - a_n \exp(-2i\lambda_n s) - (a_n - \bar{a}_n)\}\phi_n \\ \sum_{n=1}^{\infty} -\frac{d_n}{2} \{a_n \exp(2i\lambda_n s) + \bar{a}_n \exp(-2i\lambda_n s) + (a_n + \bar{a}_n)\}\phi_n \end{pmatrix}. \tag{4}$$

This formula can now be inserted into (3) to give Lp in terms of the basis ϕ_n. Since it generates a group, surjectivity of L comes down to the solvability of

$$\hat{L}p = \int_0^T p(s)e^{-s\mathscr{A}}\mathscr{B}(e^{s\mathscr{A}}u_0)\, ds = h \tag{5}$$

for $p(s)$ given $h \in X$.

4.3 An Example

We consider a vibrating beam with hinged ends and an axial load $p(t)$ as a control:

$$\begin{aligned} w_{tt} + w_{xxxx} + p(t)w_{xx} = 0, \qquad & 0 \le x \le 1 \\ w = w_{xx} = 0 \quad \text{at} \quad & x = 0, 1. \end{aligned} \tag{6}$$

Here $\lambda_n = n^2\pi^2$, $\phi_n = (1/\sqrt{2})\sin(n\pi x)$ and $d_n = -n^2\pi^2$. We can seek to solve (5) for p by expanding p in a Fourier series. For example, take $T \ge 1/\pi$ and attempt to find p's on $[0, 1/\pi]$ by writing

$$p(s) = p_0 + \Sigma\{p_{n^2} \exp(2in^2\pi^2 s) + \bar{p}_{n^2} \exp(-2in^2\pi^2 s) \tag{7}$$

and suppressing the remaining coefficients. To do this it is natural to try choosing p's in L^2. Inserting (4) and (7) into (5), we can determine h. Note that $d_n/\lambda_n = -1$ and $\{a_n \lambda_n\} \in l_2$. If we write

$$h = \begin{pmatrix} \Sigma\alpha_m\phi_m \\ \Sigma - \lambda_m\beta_m\phi_m \end{pmatrix},$$

the condition for h to be in X is $\Sigma\lambda_m^2(\alpha_m^2 + \beta_m^2) < \infty$. But the condition for h to be in the range of $\hat{L}$ with an L^2 p is that $\{a_n d_n p_{n2}\} \in l_2$. This is, however, a stronger condition than $h \in X$. Thus, we conclude that $\hat{L}$ and hence L has range that is dense in but not equal to X.

In fact, one can show that not only is L not surjective, but that (6) is *not* locally controllable in the energy norm.

To overcome this difficulty one can contemplate more sophisticated inverse function theorems, and indeed these may be necessary in general. However, for a class of equations that includes this example, a more naive trick works. Namely, instead of the X norm, make up a new space namely the range of $\hat{L}$ and use the graph norm. Miraculously, the solution $u(t, p, u_0)$ stays in this space and is still smooth in the new topology. In this stronger norm then, the implicit function theorem can still be used. The verification of these statements is somewhat lengthy, but in principle the method is straightforward.

Notes and References

1. Spaces of Solutions of Relativistic Field Theories

Problems with perturbation expansions on the flat spacetime $T^3 \times \mathbb{R}$ were first noticed by

D. Brill and S. Deser, Instability of closed spaces in general relativity, *Comm. Math. Phys.* **32** (1973), 291–304.

The terminology "linearization stability" and sufficient conditions in terms of Cauchy data were given by

A. Fischer and J. Marsden, Linearization stability of the Einstein equations, *Bull. AMS.* **79** (1973), 995–1001.

The relationship between the Cauchy data and symmetries was given by

V. Moncrief, Spacetime symmetries and linearization stability of the Einstein equations, *J. Math. Phys.* **16** (1975), 493–498.

General methods and second order conditions are given in

A. Fischer and J. Marsden, Linearization stability of non-linear partial differential equations, in *Proc. Symp. Pure Math. AMS.* **27** (1975), 219–263.

The role of the Taub conditions for linearization stability of relativity are due to Moncrief:

V. Moncrief, Spacetime symmetries and linearization stability of the Einstein equations II, *J. Math. Phys.* **17** (1976), 1893–1902.

The fact that the Taub conditions are always nontrivial conditions is proved in

J. Arms and J. Marsden, The absence of Killing fields is necessary for linearization stability of Einstein's equations, *Ind. Univ. Math. J.* **28** (1979), 119–125.

The sufficiency of the Taub conditions is proved in the following papers.

A. Fischer, J. Marsden, and V. Moncrief, The structure of the space of solutions to Einstein's equations, I One Killing field, *Ann. Inst. H. Poincaré* **33** (1980), 147–194.

A. Arms, A. Fischer, J. Marsden, and V. Moncrief, The structure of the space of solutions of Einstein's equations: II Many Killing fields, (1980), (in preparation).

The role of linearization stability in quantum gravity is explored in

V. Moncrief, Invariant states and quantized gravitational perturbations, *Phys. Rev. D*. **18** (1978), 983–989.

The appropriate general Hamiltonian formalism for studying spaces of solutions was given by

A. Fischer and J. Marsden, A new Hamiltonian structure for the dynamics of general relativity, *J. Grav. Gen. Rel.* **7** (1976), 915–920.

Results about symplectic structures on spaces of solutions may then be read off from

J. Marsden and A. Weinstein, Reduction of symplectic manifolds with symmetry, *Rep. on Math. Phys.* **5** (1974), 121–130.

Some papers dealing with gauge theories are

J. Arms, Linearization stability of the Einstein–Maxwell system, *J. Math. Phys.* **18** (1977), 830–833.
V. Moncrief, Gauge symmetries of Yang–Mills fields, *Ann. Phys.* **108** (1977), 387–400.
J. Arms, Linearization stability of gravitational and gauge fields, *J. Math. Phys.* **20** (1979), 443–453.
—, The structure of the solution set for the Yang–Mills equation (1980), preprint.

General properties of momentum maps may be found in

R. Abraham and J. Marsden, *Foundations of mechanics, Second Edition*, Addison-Wesley, 1978.

The singularities in zero sets of momentum maps are investigated in

J. Arms, J. Marsden, and V. Moncrief, Bifurcations of momentum mappings, *Comm. Math. Phys.*, **78** (1981), 455–478.

2. The Traction Problem in Nonlinear Elasticity

The work of the Italian school is represented by the following three references:

F. Stoppelli, Sull' esistenza di soluzioni delle equazioni dell' elastostatica isoterma nel case de sollectizioni dotate di assi di equilibrio, *Richerche Mat.* **6** (1957), 244–282; **7** (1958), 138–152.
G. Grioli, *Mathematical Theory of Elastic Equilibrium*, Ergebnisse der Ang. Mat. #7, Springer-Verlag, Berlin, 1962.
G. Capriz and Podio Guidugli, On Signorini's perturbation method in nonlinear elasticity, *Arch. Rat. Mech. An.* **57** (1974), 1–30.

Two general references on nonlinear elasticity relevant to the discussions in this paper are

C. Truesdell and W. Noll, *The Nonlinear Field Theories of Mechanics*, handbuch der Physik III/3, S. Flügge, Ed., Springer-Verlag, Berlin, 1965.
J. Marsden and T. Hughes, Topics in the mathematical foundations of elasticity, in *Nonlinear Analysis and Mechanics*, Vol. II, R. J. Knops, Ed., Pitman, 1978.

For the use of singularity theory in the buckling of plates, see

D. Schaeffer and M. Golubitsky, Boundary conditions and mode jumping in the buckling of a rectangular plate, *Comm. Math. Phys* **69** (1979), 209–236.

For additional details on the work described here, see

D. Chillingworth, J. Marsden, and Y. H. Wan, Symmetry and bifurcations in three dimensional elasticity I, (1981), (preprint).

3. Chaotic Oscillations of a Forced Beam

Specific experiments related to the equation (1) are discussed in these two papers:

W-Y Tseung and J. Dungundjii, Nonlinear vibrations of a buckled beam under harmonic excitation, *J. Appl. Mech.* **38** (1971), 467–476.

F. C. Moon and P. H. Holmes, A magneto-elastic strange attractor, *J. Sound and Vibration* **65** (1979), 275–296.

The Duffing equation (2) was analyzed at length by Holmes:

P. Holmes, A nonlinear oscillator with a strange attractor, *Phil. Trans. Roy. Soc.* **A292** (1979), 419–448.

—, Averaging and chaotic motions in forced oscillations, *Siam. J. on Appl. Math.* **38** (1980), 65–80.

Basic background on the Smale horseshoe is found in

S. Smale, Differentiable dynamical systems, *Bull. Am. Math. Soc.* **73** (1967), 747–817.

The original Melnikov paper is

V. K. Melnikov, On the stability of the center for time periodic perturbations, *Trans. Moscow Math. Soc.* **12** (1963), 1–57.

The detailed proofs and further discussion can be found in the following two references, especially the second:

P. Holmes and J. Marsden, Bifurcation to divergence and flutter in flow-induced oscillations; An infinite dimensional analysis, *Automatica* **14** (1978), 367–384.

——, A partial differential equation with infinitely many periodic orbits: Chaotic oscillations of a forced beam, *Arch. Rat. Mech. An.* (1981), to appear.

Background on infinite dimensional Hamiltonian systems is given in

P. Chernoff and J. Marsden, *Properties of Infinite Dimensional Hamiltonian Systems*, Springer Lecture Notes No. 425, Springer-Verlag, New York, NY, 1974.

4. A Control Problem for a Beam

Good general references for some of the current research in control theory are

D. Russell, Mathematics of finite-dimensional control systems. Theory and Design, Marcel Dekker, New York, NY, 1979.

R. Brockett, CBMS lectures on control, *SIAM*. (1980), (to appear).

The finite dimensional case involving the commutator $[\mathscr{A}, \mathscr{B}]$ can be found in, for example,

V. Jurdjevic and J. Quinn, Controllability and stability, *J. Diff. Eq.* **28** (1978), 281–289.

Some infinite dimensional results are found in

H. Hermes, Local controllability of observables in finite and infinite dimensional nonlinear control systems, *Appl. Math. and Optim.* **5** (1979), 117–125.

Results on fourier series useful for proving the range of $\hat{L}$ is dense and hence concluding controllability of any finite number of modes is found in:

J. M. Ball and M. Slemrod, Nonharmonic Fourier series and the stabilization of distributed semi-linear control systems, *Commun. Pure and Appl. Math.* **32** (1979), 555–587.

Details of the methods presented here may be found in

J. M. Ball, J. E. Marsden, and M. Slemrod, Controllability of distributed bilinear systems (1980), preprint.

Bifurcation, Catastrophe, and Turbulence

E. C. Zeeman*

Introduction

Bifurcation occurs in a parametrised dynamical system when a change in a parameter causes an equilibrium to split into two. Catastrophe occurs when the stability of an equilibrium breaks down, causing the system to jump into another state. The elementary theory concerns dynamical systems with steady state equilibria (point attractors), and the non-elementary theory concerns systems with dynamic equilibria (periodic attractors and strange attractors). In the elementary case Thom [72] has used singularities to classify both bifurcation and catastrophe, and this has led to a great variety of applications [22]. We illustrate the contrasting styles of application in biology and physics by describing two recent examples. The first is a model by Seif [59] concerning hyperthyroidism, and the second is a model by Schaeffer [58] concerning Taylor cells in fluid flow.

In the non-elementary theory there is no classification yet, but Ruelle and Takens [51] have suggested using strange attractors to model the onset of turbulence. We describe some strange attractors and strange bifurcations, and discuss some of their properties that resemble turbulence, such as stability yet sensitive dependence on initial condition, and the broad band frequency spectra similar to those observed by Swinney and Gollub [67]. It is a pleasure to acknowledge my debt to David Rand for many discussions.

1. Elementary Examples

In this section we illustrate the difference between bifurcation and catastrophe by two simple examples and explain the relationship between them

* Department of Mathematics, University of Warwick, Coventry, England.

by a third example containing the first two. We draw attention to the important properties of stability, locality and symmetry. In the following two sections we shall give definitions, and explain how these simple examples typify the situation in higher dimensions.

1.1. The Pitchfork Bifurcation. This is the simplest example of a bifurcation, and has a canonical equation

$$\dot{x} = bx - x^3.$$

Here x is a real variable and b is a real parameter; $\dot{x}$ denotes dx/dt. The equilibrium set M is given by $bx - x^3 = 0$, and is shown in Figure 1. For

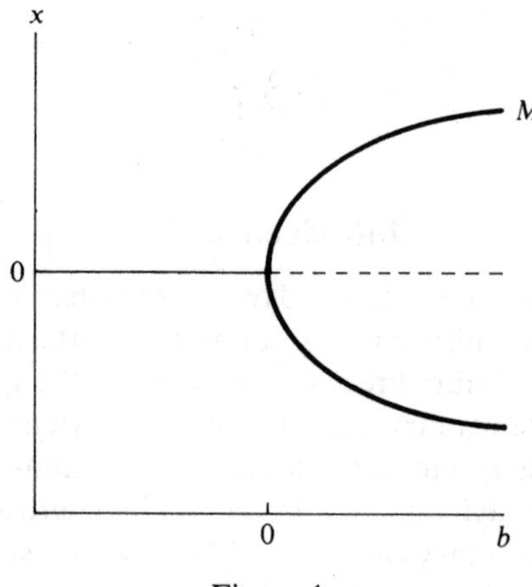

Figure 1

$b < 0$ there is a unique attractor (or stable equilibrium) at $x = 0$; for $b > 0$ there are two attractors at $x = \pm\sqrt{b}$, separated by a repellor (or unstable equilibrium) at $x = 0$. In Figure 1 the repellors are shown dotted. Therefore if the parameter is increased from negative to positive the attractor bifurcates at $b = 0$.

The pitchfork is unstable and local, as we shall now explain. It is *unstable* because there exist arbitrarily small perturbations with a topologically different equilibrium set (see Definition 2.2 below). For example the perturbation

$$\dot{x} = \varepsilon + bx - x^3, \qquad \varepsilon > 0,$$

breaks M into two components as shown in Figure 2. The component containing the unique attractors when $b < 0$ is called the *primary branch* because if b is increased from negative to positive the system will follow this branch. The attractors on the primary branch are called *primary modes*, and those on the other component are called *secondary modes*. The instability of the pitchfork can be analysed by unfolding it; here *unfolding* means adding

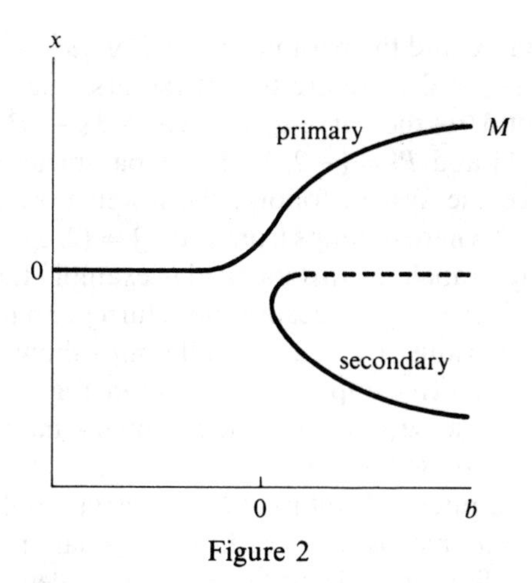

Figure 2

more parameters to make it stable. We shall see in 1.3 below that one more parameter is sufficient.

Meanwhile the pitchfork is *local* because it possesses an organising centre; here an *organising centre* means a point where the local dynamics is equivalent to the global dynamics of the whole system (see definition 2.3). The advantage is that the global dynamics can then be analysed locally at the organising centre. In the pitchfork the organising centre is the origin $x = b = 0$.

1.2. The Catastrophic Jump. The catastrophic jump PQ shown in Figure 3 has a beginning P and an end Q. The beginning P is a fold catastrophe, where an attractor and a repellor coalesce and disappear. The end Q is another attractor. The dynamical system illustrated in Figure 3 has the equation

$$\dot{x} = a + 3x - x^3,$$

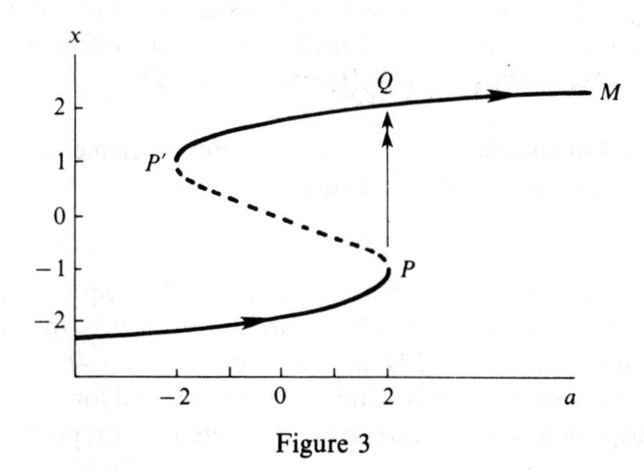

Figure 3

where the variable is x, and the parameter is a. For $|a| > 2$ there is a unique attractor, and for $|a| < 2$ there are two attractors separated by a repellor. The equilibrium set M is the curve given by $a + 3x - x^3 = 0$, and this has folds at $P = (2, -1)$ and $P' = (-2, 1)$. If the parameter is increased from negative to positive the system follows the lower attracting branch of M until it reaches $a = 2$ where it jumps from P to $Q = (2, 2)$, and it then follows the upper attracting branch. In this particular example there is hysteresis if the parameter is reduced again, because the return jump takes place at P', at a different parameter value $a = -2$. Not all jumps show hysteresis because there may not be a return jump; for instance, in Figure 2 a decrease in b causes a jump from the secondary to the primary mode, but there is no return jump if b is increased again.

In contrast to the pitchfork bifurcation the catastrophic jump is stable and non-local. *Stable* means that sufficiently small perturbations have equivalent M (see definition 2.2) and hence an equivalent jump. Meanwhile it is *non-local* because any analysis of the dynamics must involve both the taking-off point P and the landing point Q, and so there is no organising centre for the whole jump. It is true that the fold catastrophe by itself is local because it has an organising centre at P, but this would be an incomplete analysis of the global jump PQ.

Remark. The difference between local and global helps to explain why bifurcation theory is older than catastrophe theory. A local system can be linearised at the organising centre and treated as an eigenvalue problem; it can be analysed both quantitatively and qualitatively by examining the higher order terms of the Taylor expansion at the organising centre, using the methods of classical analysis. By contrast a non-local system may have to be tackled by methods of topology or modern global analysis, and may yield only qualitative rather than quantitative properties. Thus catastrophe theory had to wait for the development of twentieth-century topology, whereas bifurcation theory already had at its disposal nineteenth-century analysis.

On the other hand some non-local systems can be localised by identifying them with sections of a higher dimensional local system, with a hidden organising centre, as illustrated by the next example.

1.3. The Cusp Catastrophe. This is the next simplest elementary catastrophe after the fold, and has a canonical equation

$$\dot{x} = a + bx - x^3,$$

where x is the variable, and a, b are parameters. The equilibrium set is the surface M shown in Figure 4. The bifurcation set B is the image in the parameter space of the folds of M, and is the cusp $27a^2 = 4b^3$. For parameter points outside the cusp there is a unique attractor, and for parameter points inside the cusp there are two attractors separated by a repellor. In Figure 4

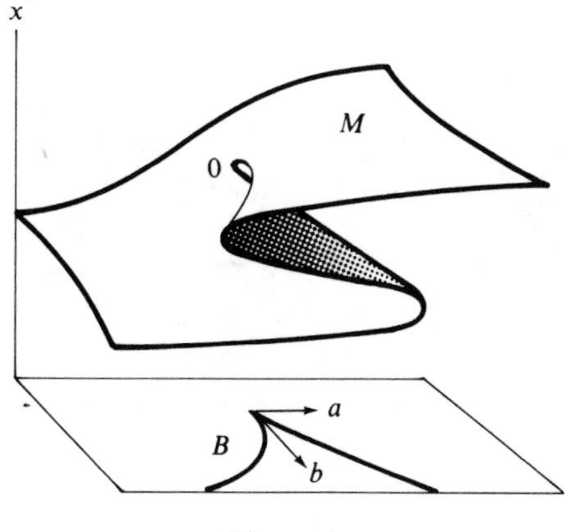

Figure 4

the shaded middle sheet repels, while the rest of M attracts. In applications a is sometimes called a *normal* factor, because it is correlated with x, and b is called a *splitting* factor, because it splits the attractor surface apart.

The cusp catastrophe is both stable and local, and contains both the previous examples as sections. The pitchfork in Figure 1 is the section $a = 0$, and hence the cusp catastrophe is its unique unfolding.* The perturbation in Figure 2 is the section $a = \varepsilon$. The jump in Figure 3 is the section $b = 3$. The origin 0 is the organising centre, and any section through 0 is unstable and local, like the pitchfork. Meanwhile, any section not through 0 that cuts the cusp transversally is stable and non-local because it contains a jump; we call 0 the *hidden* organising centre of these sections. Summarising: the local analysis of the cusp catastrophe at 0 elucidates the global dynamics of both pitchfork and jump, and reveals their interrelationship.

1.4. Symmetry. The pitchfork is symmetrical with respect to change of sign of x, but the cusp is not. We therefore call the b-parameter *symmetric* and the a-parameter *asymmetric*. Similarly any linear combination of a, b is asymmetric, and so the pitchfolk is the maximal symmetric section of the cusp.

In some problems the idealised model is symmetric, while the real model is represented by an asymmetric perturbation. For example the ideal elastic beam can be modelled by a pitchfork with b representing compression and x the resulting buckling [76]. The ideal beam is symmetric because it can buckle equally well up or down, but a real beam is liable to contain imperfections that make it behave like the asymmetric perturbation in Figure 2; hence it will always buckle the same way when compressed because it has to follow the primary branch. Summarising the procedure for such problems: represent the idealised problem by a symmetric bifurcation, unfold it, take

* An unfolding is unique up to equivalence [33, 72, 76]. Golubitsky and Schaeffer [16, 17] obtain a 1-dimensional higher unfolding because they retain b as a distinguished parameter.

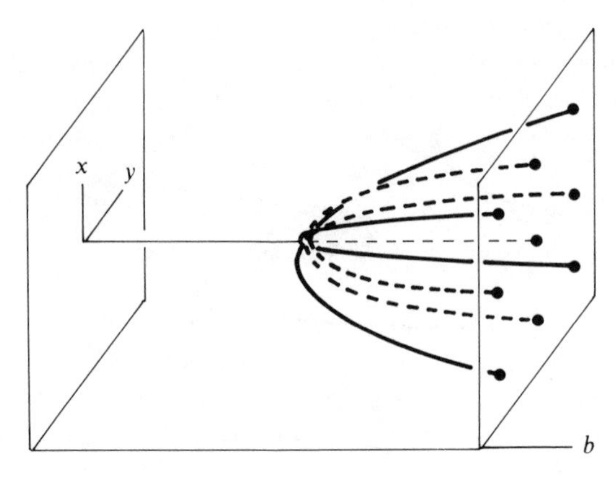

Figure 5

the maximal symmetric section, perturb it in an asymmetric direction, and then take the primary branch.

Of course, in the above special case the pitchfork turned out to be the same as the maximal symmetric section of its unfolding, but in the general case the latter may be larger. For example, consider the double-pitchfork shown in Figure 5 and given by the equations

$$\dot{y} = by - y^3$$

$$\dot{z} = bz - z^3$$

where y, z are variables and b is a parameter. Notice that b is a symmetric parameter with respect to changes of sign of both y and z. The double-pitchfork unfolds to the double-cusp with 8 parameters, of which 3 are symmetric and 5 asymmetric. One of the symmetric parameters is modal, and can be ignored for technical reasons [5, 10, 17, 27, 76]. Therefore the maximal symmetric section is given by adding the other symmetric parameter a, which has the effect of pulling the two pitchforks apart along the b-axis:

$$\dot{y} = (b + a)y - y^3$$

$$\dot{z} = (b - a)z - z^3.$$

Perturbing in the direction of an asymmetric parameter c and taking the primary branch will give a stable non-local equilibrium surface M over the (a, b)-plane, whose shape will depend on c. An example is shown in Figure 13 below.

2. Stability and Organising Centres

In this section we give the definitions of stability and organising centre for general flows, and in the next section we specialise to the elementary theory.

Let X be a manifold and ϕ a flow on X. Here a *flow* means a C^∞-map

$\phi: \mathbb{R} \times X \to X$, $\phi(t, x) = \phi^t x$, such that $\{\phi^t\}$ is a group action of $\mathbb{R}$ on X. The ϕ-*orbit* through x is $\phi(\mathbb{R} \times x)$. ϕ is the solution of a differential equation on X.

2.1. Definition of Nonwandering set. Call a point $x \in X$ *nonwandering* if, $\forall$ neighbourhood N of x, $\forall t \in \mathbb{R}$, $\exists s > t$, such that $\phi^s N$ meets N. The set of all such is called the *nonwandering set* Ω. Let M denote the *fixed point set* (or equilibrium set) of ϕ. In the elementary case $\Omega = M$, but in the general case $\Omega \supset M$ because Ω will also contain periodic orbits, strange attractors, etc. (See Sections 6 and 7 for examples.) Since all orbits flow to Ω the asymptotic behaviour of ϕ is determined by Ω.

Parameters. Let C be a parameter manifold. Let ϕ be a flow on X parametrised by C; here ϕ is a C^∞-map $\phi: C \times \mathbb{R} \times X \to X$, $\phi(c, t, x) = \phi_c(t, x)$, and ϕ_c is a flow on X, $\forall c \in C$. Define the parametrised nonwandering set

$$\Omega = \bigcup (c \times \Omega_c) \subset C \times X,$$

where Ω_c is the nonwandering set of ϕ_c, $\forall c \in C$. Let $\chi: \Omega \to C$ be induced by the projection $\pi: C \times X \to C$. For example, in the cusp catastrophe (1.3 above) $\Omega = M$, the parametrised fixed point set, and χ is the projection onto $\mathbb{R}^2$ shown in Figure 4.

2.2. Definition of Stability. Given flows ϕ, ϕ' on X, X' parametrised by C, C' with nonwandering sets Ω, Ω' define them to be *equivalent*, written $\phi \sim \phi'$, if $\exists$ a homeomorphism α throwing orbits to orbits, and a diffeomorphism γ, such that the following diagram commutes:

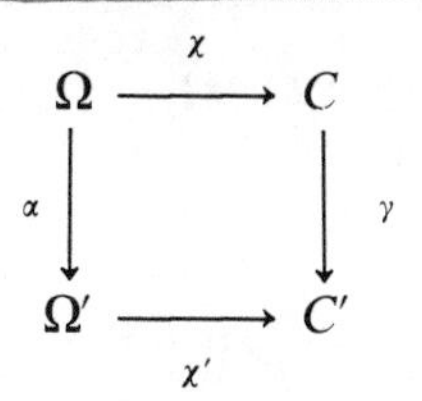

Define ϕ to be *stable* if perturbations are equivalent. Here a *perturbation* of ϕ means the solution of a differential equation sufficiently close to that for ϕ.

Remark. The above definition is usually called Ω-stability to distinguish it from structural stability, in which the homeomorphism α and the commutativity extend to the whole of $C \times X$. Structural stability was first introduced by Andronov and Pontryagin [3], and the weaker definition of Ω-stability was later introduced by Smale [64] to embrace a larger class of flows. Neither definition is generic, because Ω-stable flows are not dense [2]. It was originally hoped to find a generic definition, but hopes dwindled as more examples were discovered. Meanwhile the importance of structural stability and Ω-stability was emphasised by theorems characterising them [39, 44, 48, 49, 65]. For simplicity of exposition we have chosen Ω-stability in this paper.

In elementary theory the situation is much tidier because the definition is generic, and for low dimensions of C one can sharpen α to be a diffeomorphism, as we explain in the next section. For more detailed discussions of stability see [1, 30, 64, 72, 73].

2.3. Definition of Organising Centre. Given a flow ϕ on X parametrised by C we call $P \in \Omega$ an *organising centre* for ϕ if $\exists$ arbitrarily small neighbourhoods Ω', C' of P, χP in Ω, C, a homeomorphism α throwing orbits to orbits, and a diffeomorphism γ, such that the following diagram commutes:

$$\begin{array}{ccc} \Omega', P & \xrightarrow{\chi|\Omega'} & C' \\ \downarrow \alpha & & \downarrow \gamma \\ \Omega, P & \xrightarrow[\chi]{} & C \end{array}$$

Call ϕ *local* if it has an organising centre.

2.4. Lemma. *The cusp catastrophe is stable and local.*

PROOF. For the stability see [76]; we verify here that it is local. The cusp is defined in 1.3 above and has $X = \mathbb{R}$, $C = \mathbb{R}^2$, and $\Omega = M \subset C \times X$ given by $a + bx - x^3 = 0$. Given $\varepsilon > 0$, let C' be the open disk $a^2 + b^2 < \varepsilon$, and let $M' = \chi^{-1}C'$. Define $\alpha: M' \to M$, $\gamma: C' \to C$ by

$$\alpha(a, b, x) = (s^3 a, s^2 b, sx)$$
$$\gamma(a, b) = (s^3 a, s^2 b), \qquad s = \sec \frac{\pi}{2\varepsilon}(a^2 + b^2).$$

This satisfies commutativity, and hence the origin is an organising centre. Notice that since this example is elementary α is not only a homeomorphism but also a diffeomorphism, and it extends to the ambient space.

2.5. Definition of Bifurcation Set and Catastrophe Set. Given a flow ϕ on X parametrised by C, let Ω_* denote the map $c \to \Omega_c$ from C to the space of closed sets of X, with the Hausdorff topology. Call a parameter point c *regular* if Ω_* is constant in a neighbourhood of c, up to equivalence. Define the *bifurcation set* B to be the set of non-regular points. Define the *catastrophe set* K to be the closure of the set where Ω_* is discontinuous. Therefore we have closed subsets

$$K \subset B \subset C.$$

A point in K is called a *catastrophe point* and a point in $B - K$ is called a *bifurcation point*. Thus Ω_* is continuous in the neighbourhood of a bifurcation point, but discontinuous in the neighbourhood of a catastrophe point.

For example, the pitchfork Ω_* is continuous, so $K = \varnothing$ and the organising centre is a bifurcation point; that is why it is called the pitchfork bifurcation. By contrast, in the cusp catastrophe Ω_* is continuous at the organising centre but discontinuous elsewhere on the cusp, so $K = B$ and the organising centre is a catastrophe point; that is why it is called the cusp catastrophe rather than the cusp bifurcation. For a philosophical discussion see Thom [72].

3. Elementary Theory

In this section we explain Thom's density and classification theorem for elementary catastrophes.

3.1. Definition of Elementary. Given a flow ϕ on X, call ϕ *elementary* if there exists a Lyapunov function. Here a *Lyapunov function* means a C^∞-function $f: X \to \mathbb{R}$ that decreases strictly along orbits of ϕ. If ϕ is elementary the nonwandering set of ϕ is the critical set of f, given by $df = 0$, and the attractors of ϕ are the minima of f.

We now introduce parameters. Given a flow ϕ on X parametrised by C, call ϕ elementary if there exists a parametrised Lyapunov function f. Here f is a C^∞-function $f: C \times X \to \mathbb{R}$, $f(c, x) = f_c x$, such that f_c is a Lyapunov function for ϕ_c, $\forall c \in C$. Therefore

$$\Omega = M = \text{critical set of } f, \quad \text{given by } d_X f = 0.$$

Examples. Gradient flows are elementary; here a *gradient flow* means the solution of a gradient differential equation $\dot{x} = -\nabla f$, and this is elementary because f is a Lyapunov function. For instance, the cusp catastrophe is elementary because it is gradient:

$$\dot{x} = -\nabla_X f = -\frac{\partial f}{\partial x}, \quad \text{where } f = \tfrac{1}{4}x^4 - \tfrac{1}{2}bx^2 - ax.$$

Therefore $\Omega = M$ is given by

$$d_X f = \frac{\partial f}{\partial x} = x^3 - bx - a = 0.$$

Gradient-like flows are also elementary [64]. On the other hand, the examples in Sections 6 and 7 below are non-elementary because they contain periodic orbits.

Remark. The restriction to elementary theory is a severe restriction, but nevertheless there are many phenomena in which steady-state equilibria are predominant, and so the elementary theory does have wide application [15, 21, 22, 43, 72, 76]. In the elementary theory the asymptotic behaviour of the flow ϕ is determined by the critical set M of its Lyapunov function f, and so the mathematical trick is to switch attention from ϕ to f. This trick does have some limitations [19] but is adequate for most applications. The mathematical advantage of switching to functions is that we can sharpen the definition of stability and use results from singularity theory, as follows.

3.2. Definition of Stability of Functions. Given functions f, f' on X, X' parametrised by C, C' define them to be *equivalent*, written $f \sim f'$, if $\exists$ diffeomorphisms α, β, γ such that the diagram commutes:

$$\begin{array}{ccccc} C \times X & \xrightarrow{\pi \times f} & C \times \mathbb{R} & \xrightarrow{\pi} & C \\ \downarrow{\scriptstyle \alpha} & & \downarrow{\scriptstyle \beta} & & \downarrow{\scriptstyle \gamma} \\ C' \times X' & \xrightarrow{\pi \times f'} & C' \times \mathbb{R} & \xrightarrow{\pi} & C' \end{array}$$

This induces an equivalence of critical sets:

$$\begin{array}{ccc} M & \xrightarrow{\chi} & C \\ \downarrow{\scriptstyle \alpha|M} & & \downarrow{\scriptstyle \gamma} \\ M' & \xrightarrow{\chi'} & C' \end{array}$$

Therefore if f, f' are Lyapunov functions for the flows ϕ, ϕ' then

$$f \sim f' \Rightarrow \phi \sim \phi'.$$

Let $F = C^\infty(C \times X)$, the space of all C^∞-functions on X parametrised by C, with the Whitney C^∞-topology [76]. Given $f \in F$ call f *stable* if it has a neighbourhood of equivalents. If f is stable then its critical set M is a manifold the same dimension as C, and $\alpha | M$ above is a diffeomorphism.

Example. Morse functions are stable; here a *Morse function* means a function on $\mathbb{R}^n$ that is quadratic of rank n, and independent of the parameter C.

The cusp catastrophe has a stable Lyapunov function, $f = x^4 - 2bx^2 - 4ax$ (multiplying by 4 to get rid of fractions). The germ of f at the organising centre is x^4, and, conversely, f is the unique unfolding of its germ, up to equivalence. Here *unfolding* means stabilising by adding the minimum number of parameters. In this sense the cusp is uniquely determined by its germ. (For proofs see, for example, [76, Chapter 18].)

Analogous to the cusp there are 11 elementary catastrophes* for $\dim C \leq 5$, determined by the germs x^3, x^4, x^5, x^6, x^7, $x^2y \pm y^3$, $x^2y + y^4$, $x^2y \pm y^5$, $x^3 + y^4$, and each one is stable and local. Moreover, this is a complete classification in the sense of Theorem 3.4 below. In order to state the classification theorem globally it is necessary to localise the definitions of equivalence and stability, as follows.

3.3. Definition of Local Stability. Given functions f, f' on X, X' parametrised by C, C', and points y, y' in $C \times X$, $C' \times X'$, define them to be *locally-equivalent*, written $(f, y) \sim (f', y')$, if $\exists$ neighbourhoods N, N' of y, y' such that $f|N \sim f|N'$. Given $f \in F$, where $F = C^\infty(C \times X)$, call f *locally-stable* if, $\forall y \in C \times X$, $\forall$ neighbourhood N of y, $\exists$ a neighbourhood V of f, $\forall f' \in V$, $\exists y' \in N$, such that $(f, y) \sim (f', y')$. Notice that stable implies locally-stable, but not conversely. Local-stability implies that the critical set M is a manifold, of the same dimension at C.

3.4. Theorem (Thom [72] and Mather [33]). *If* $\dim C \leq 5$ *then locally-stable functions are dense in* F. *If* f *is locally-stable then at each point it is locally-equivalent to either a linear function, or a Morse function, or the product of an elementary catastrophe with a Morse function.*

For a proof see, for example, [31, 76 Chapter 18].

This theorem explains why the elementary catastrophes and their sections are so universal. As in 1.3, each elementary catastrophe has two types of section, depending upon whether or not the section goes through the organising centre. If it does then the section gives an unstable local bifurcation like the pitchfork. If it does not then the section will generically give a stable non-local system like the jump, consisting of a configuration of lower dimensional catastrophes. This configuration is determined by the hidden organising centre, hidden because it lies off the section. When such a configuration is observed it may be useful to look for a hidden organising centre, because the latter may provide an analysis of the dynamics, as illustrated in 5.4 below. Summarising: the identification of pitchfork and jump as sections of the cusp typifies the relationship between bifurcation and catastrophe in higher dimensions.

In the next two sections we illustrate the use of elementary catastrophe theory by describing two recent applications. The first is in medicine, and typifies applications in the biological or social sciences where the catastrophe is taken as a hypothesis. The second is in fluid mechanics, and typifies applications in the physical sciences where the singularity is derived from basic underlying assumptions, such as conservation laws or variational principles, etc.

* Thom's list of elementary catastrophes [72, 76] has been considerably extended to higher dimensions by Arnold [5], but the definitions have to be modified to allow for the appearance of modal parameters.

4. Hyperthyroidism

This is a model by Seif [59]. In normal individuals there is a hormone* chain:

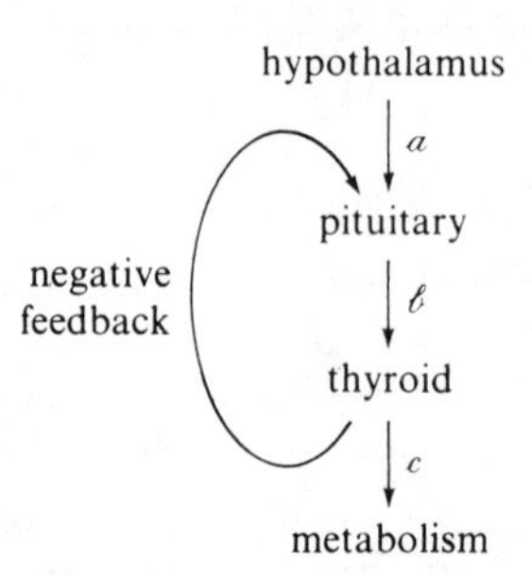

For instance, a sensation of cold would be registered in the brain by activity of the hypothalamus, causing a release of hormone a into the blood stream, which triggers the pituitary gland to release hormone b, which triggers the thyroid gland to release hormone c, which stimulates the metabolism throughout the body, changing chemical energy into heat. The negative feedback means that large c (in other words, a high concentration of c in the blood) causes the pituitary to stop releasing b. There are two ways this chain can go wrong, hypo and hyper, as follows.

4.1. Hypothyroidism. Hypothyroidism means too little c. This in turn causes large b due to lack of negative feedback. Also, every time the hypothalamus receives a stimulus and releases a in an attempt to get the metabolism going nothing happens, and so it continues to release a, reinforcing the large b. The typical cause of too little c is a lack of iodine, without which the thyroid cannot manufacture c. Typical symptoms of hypothyroidism are feelings of cold and sluggishness, and a tendency for the individual to grow fat. A cure is to inject c.

4.2. Hyperthyroidism. Hyperthyroidism means too much c. This in turn causes small b, due to the feedback. The typical cause is the presence of some immunoglobulin b^* in the blood that accidentally resembles b, and which the thyroid mistakes for b. Since b^* is always present in the blood the thyroid is permanently turned on to the production and release of c. The original appearance of b^* may have been triggered by some quite independent immunological response, but once triggered it is always present, thus inducing a permanent state of hyperthyroidism. Typical symptoms of hyperthyroidism are that the individual feels hot and irritable, becomes overactive and thin, develops bulbous eyeballs and a bulge in the neck called a goitre

* a = TRH(thyrotropic releasing hormone) = protirelin.
b = TSH(thyroid stimulating hormone) = thyrotropin.
c = index measuring thyroxine (T_4) and triiodothyronine (T_3).

due to the enlarged overactive thyroid. A temporary treatment is to inject drugs that interfere with the metabolism, thus counteracting the effect of large c. The permanent treatment is to remove or destroy some of the thyroid gland until what is left produces the desired amount of c.

Seif applied this standard treatment to 78 hyperthyroid patients, but found that although the treatment produced normal levels of b and c, and indeed cured two-thirds of his patients, the other third were apparently not properly cured because they began to display some of the opposite symptoms of hypothyroidism, such as the inability to react to cold.

4.3. Pituitary Failure. What had happened was that the pituitary was no longer responding to the a-hormone; evidently during the long period of high negative feedback, in a desperate attempt to stem the flood of c, the pituitary had given up releasing b, and consequently lost its response-ability. Seif measured this response-ability by injecting a standard dose of a and observing the resulting change in the level of b. Let $x = b_1/b_0$, where b_0 is the initial level, and b_1 the level twenty minutes after injection, which is the normal time for maximal response. He found that for normal individuals and for the successfully cured two-thirds x had mean 7, but for hyperthyroid patients and for the abnormal third of treated patients $x = 1$. For convenience we shall refer to the abnormal third by the single word *treated*. Figure 6 sketches the four types on the (b, c)-plane, and Figure 7 sketches their response to the injection of a.

Seif had collected 422 measurements during the diagnosis and treatment of 314 patients, but at this point he was stuck for a conceptually simple way to present, or think about, his data. That is until he happened to hear a lecture mentioning the cusp catastrophe, which he immediately recognised as relevant, because x was bimodal for some points in the (b, c)-plane but unimodal for others. The resulting model is shown in Figure 8.

4.4. Cure. The diagram immediately suggested the possibility of catastrophic jumps α and β. The jump α represents the onset of hyperthyroidism, and the jump β represents a possible cure for the treated patients. Their treatment so far is represented by the path HK in the parameter space, which

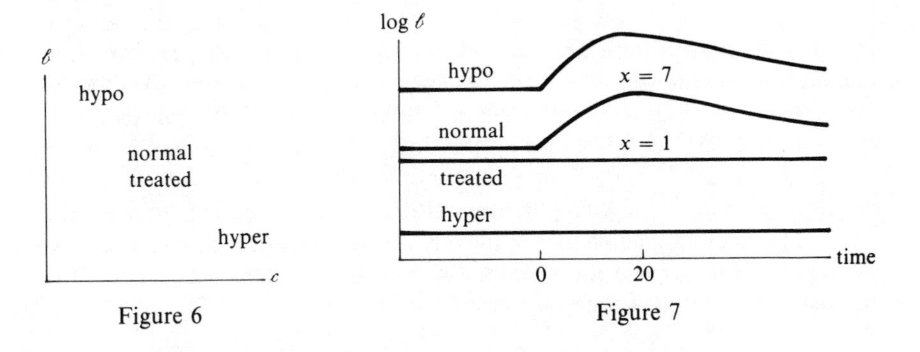

Figure 6

Figure 7

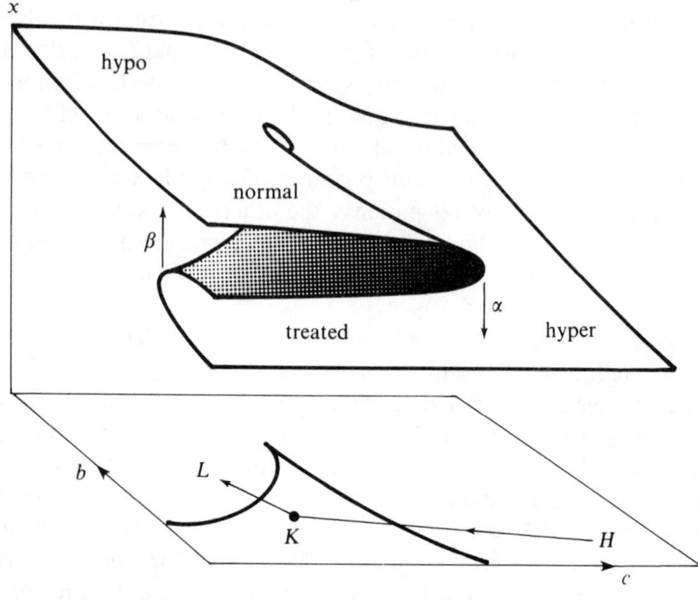

Figure 8

takes about 6 weeks, and is the slow change in the average levels of b, c resulting from the removal of some of the thyroid gland. The additional treatment would be to steer the patient along the path KL, until the left side of the cusp is crossed, where the patient will jump back to normal. This guiding of the patient through hypothyroidism takes about another 3 weeks and is achieved by drugs that stimulate the production of a by reducing the metabolism, and suppress the negative feedback by interfering with the production of c, thus inducing the pituitary to very gradually raise the average level of b. To Seif's delight this new additional treatment successfully cured the remaining third of his patients. Evidently the other two-thirds had cured themselves either by already crossing the left side of the cusp or by going round the top.

It could be argued that, having identified the pituitary as the culprit, the guiding of the patient through hypothyroidism would have been a logical way to try and cure the patient anyway, without reference to Figure 8. However without Figure 8 one would not have predicted the suddenness of the jump β, nor the stability of the patient's normality immediately after the jump. Therefore Figure 8 makes the monitoring of the additional treatment conceptually very easy, because all the doctor has to do is to make regular injection tests until he observes a jump in x, and he can then send the patient home cured. Going back to our original discussion at the beginning of the paper, of the relationship between the pitchfork and the jump, what was first observed in this application was the bifurcation, and what was successfully predicted as a result was the jump.

4.5. Data Fitting. The next step was to fit a cusp to the data. Theoretically (Theorem 3.4) the cusp surface is only differentially equivalent (Definition 3.2) to the standard surface (Example 1.3), but near the cusp point a linear fit is a good approximation. Therefore Seif assumed that, up to additive and multiplicative constants, the variable was log x, the normal factor was c, and the splitting factor $-b$; he calculated the constants by a least squares fit on x using an iterative computer programme. The resulting cusp lines are shown in Figure 9; the curvature of the right side is reversed due to the log-scale. Notice that, as anticipated from Figure 8, normal individuals are bounded on the right by the right side of the cusp, and the treated patients are bounded on the left by the left side of the cusp.

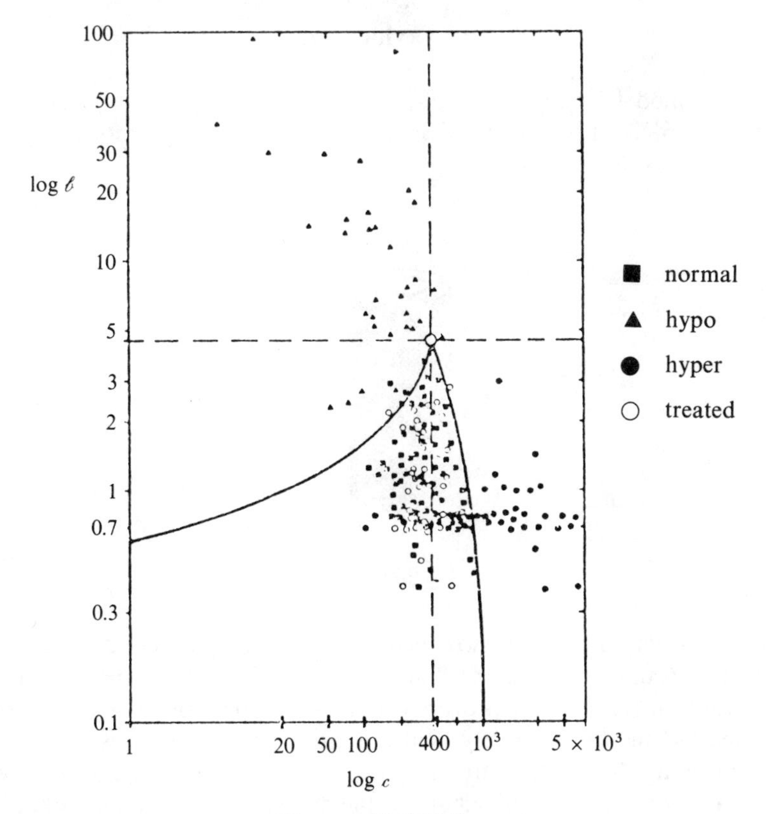

Figure 9 (Seif [59]).

4.6. Micro-Model. The next task was to explain the cusp, because a good model should not only provide a structural description at the phenomenological level, but also admit a reductionist explanation at the micro-level. Therefore Seif constructed a stochastic microscopic model of granule formation, stimulus-secretion, and diffusion, of the b-hormone in individual pituitary cells. He then showed that the resulting macroscopic behaviour of the pituitary gland would have an equilibrium surface with the same formula that he had used for the data fitting. Therefore Figure 8 represents not only a distribution of patients, but also the behaviour of the pituitary gland.

4.7. Summarising. Seif's model is useful in five ways. Firstly, it provided a simple conceptual grasp of the problem. Secondly, it suggested a successful cure. Thirdly, it enabled the data to be fitted. Fourthly, it stimulated the construction of a micro-model, and provided an objective for the latter to explain. Fifthly, it provides a theoretical framework for ongoing research into the molecular substructure.

5. Taylor Cells

This is a model by Schaeffer [58] to explain the experimental results of Benjamin [6] in the classical Couette [11]–Taylor [71] problem in fluid mechanics (see Figure 10).

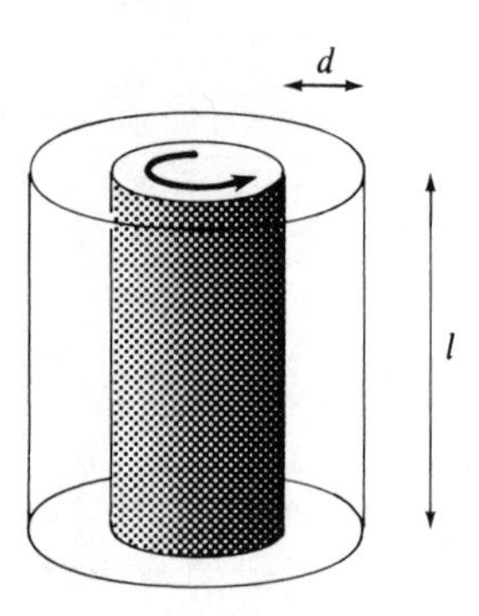

Figure 10

Water is placed between two vertical concentric cylinders, and the inner cylinder is rotated causing the fluid to rotate. We take as a parameter the Reynolds number R, which is proportional to the speed of the inner cylinder. For small R the fluid velocity field ξ is azimuthal, in other words all the streamlines are horizontal circles concentric with the cylinders. If the end effects are ignored (or equivalently if the cylinders are infinitely long) it is easy to calculate ξ from the Navier–Stokes equations [26]; ξ depends on the cylinder speed but turns out to be independent of viscosity, and is called Couette flow [11].

If R is increased then Couette flow becomes unstable; cells appear, called Taylor cells, and the fluid settles down into a new type of steady flow called Taylor flow [71]. Each cell is a horizontal solid torus of approximately square cross-section. Inside each cell the flow is spiral; there is one horizontal circular streamline in the interior of the cell and all the other streamlines spiral round it. Alternate cells spiral clockwise and anticlockwise, as shown in Figure 11. The boundaries between the adjacent cells are horizontal, and alternate boundaries spiral inwards and outwards. Intuitively it is the outward spiralling boundaries that are being driven by centrifugal force, and they in turn drive the cells.

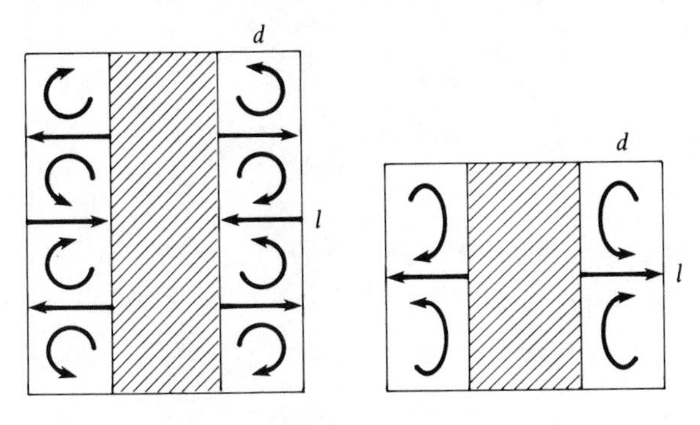

Figure 11

We now want to take the end effects into account. Define the *aspect ratio* to be $\rho = l/d$, where l is the length of the cylinders and d the clearance between them. Since the cells are approximately square in cross-section, the number of cells is approximately equal to ρ. At the ends of the cylinders the velocity is zero due to friction, and so in the boundary layers at the ends the velocity is small; therefore the centrifugal force is small, and this biases the end layers to spiral inwards. This bias normally causes an even number of cells to form, as in Figure 11. If, however, we make $\rho = 3$, then the poor fluid does not know whether to form 2 cells or 4 cells, and this is the phenomenon we want to discuss.

5.1. Experimental Data. Benjamin [6] performed the experiment, taking as parameters the Reynolds number R and the aspect ratio ρ. He found that if ρ is fixed and R varied then at certain critical values of R the fluid jumps from 2 cells to 4 cells or vice versa. By "jump" we mean that on one side of the critical value both configurations of cells are stable, but if R is moved across the critical value then one of the configurations loses its stability, and the fluid will settle down into the other; the time it takes to jump will depend upon the viscosity. Let C denote the region of the parameter plane in which the experiment is valid. Benjamin plotted the observed jump points in C and obtained the cusp-shaped bifurcation curve in Figure 12.

Theoretically, if we choose a suitable variable v and plot v over C then we should obtain the cusp catastrophe surface M shown in Figure 13. For example, let v be the inward radial component of the velocity at a point halfway up the cylinders midway between them. This is a suitable variable because $v < 0$ in the 2-cell flow, $v = 0$ in the Couette flow, and $v > 0$ in the

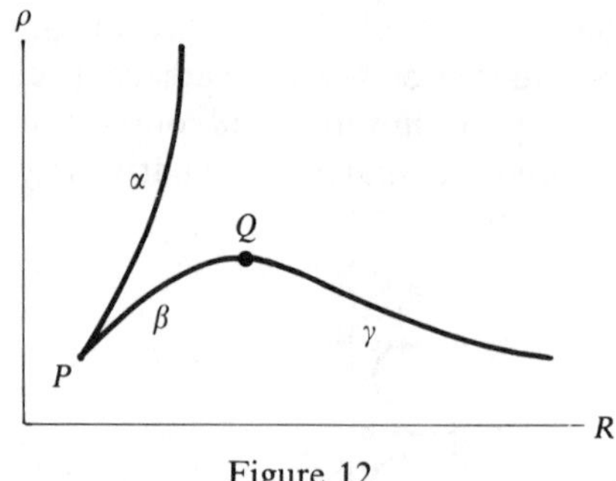

Figure 12

4-cell flow, as can be seen from Figure 11. Therefore for large R the lower sheet of M in Figure 13 represents the 2-cell mode and the upper sheet the 4-cell mode.

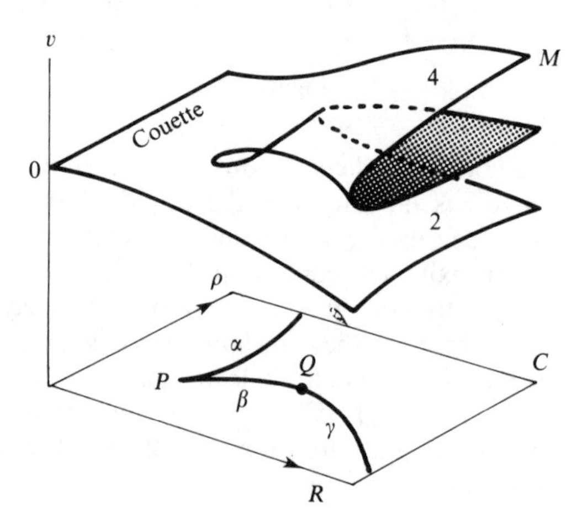

Figure 13

5.2. Digression on the Meaning of "Suitable." Let X be the ∞-dimensional space of all possible fluid velocity fields in the apparatus. The Navier–Stokes equations determine an evolution equation E on X, parametrised by C. The equilibrium set of E is a cusp catastrophe surface $M \subset C \times X$. Let $v\colon X \to \mathbb{R}$ be a function, for example some measurement of the velocity field. We call v a *suitable* variable if the composition

$$M \xrightarrow{\subset} C \times X \xrightarrow{1 \times v} C \times \mathbb{R}$$

is an embedding. Notice that $1 \times v$ crushes the ∞-dimensional space $C \times X$ down onto the 3-dimensional space $C \times \mathbb{R}$, but does not crush M. Notice also that the dynamic E sits up in $C \times X$, and that there is no dynamic in $C \times \mathbb{R}$, only the equilibrium set M and the catastrophic jumps, as follows.

5.3. Qualitative Description. The two important points of the bifurcation set are the cusp point P, and the point Q where the tangent is parallel to the R-axis. These points divide the bifurcation set into three arcs α, β, γ as shown in Figure 12. The reason why these points are important is that in each experiment ρ is fixed and so the variation of R is represented by a section $\rho = \text{constant}$; all such sections are stable except those through P and Q, where the equivalence class of section changes, as shown in Figure 14. If

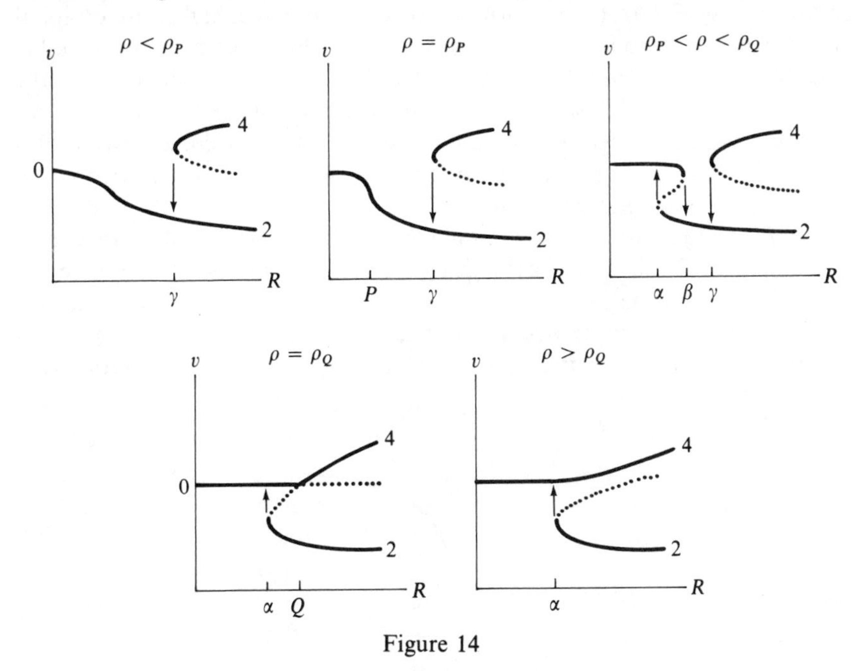

Figure 14

$\rho < \rho_P$ then the 2-cell mode is primary and the 4-cell mode secondary; if $\rho > \rho_Q$ the situation is reversed. Unstable modes are shown dotted. If $\rho_P < \rho < \rho_Q$ there is a hysteresis on the primary branch between an incipient 4-cell mode and the 2-cell mode, while the 4-cell mode is secondary. If the system is in the 2-cell mode and R is decreased across the arc α then it will jump into the 4-cell mode. The reverse jump occurs if R is increased across β or decreased across γ. A similar phenomenon happens in embryology [76].

5.4. Mathematical Analysis. Benjamin was unable to solve the Navier–Stokes equations because of the difficulties presented by the boundary conditions at the ends of the cylinders. Taylor's original solution [76] had avoided this difficulty by assuming infinitely long cylinders and proceeding as follows. For small R the Couette flow ξ is the unique attractor of the evolution equation E on X, and so he used (R, ξ) as an organising centre; by linearising E at the organising centre he was able to calculate the critical Reynolds number R_c where ξ becomes unstable, and express the Taylor flow as a perturbation of ξ. The reason that Benjamin was unable to follow the same procedure was that Figure 13 is non-local since there are two qualitatively significant points P and Q.

Schaeffer [58] realised this and had the brilliant idea of seeking a *hidden* organising centre. We sketch his procedure briefly, as follows. His first trick was to introduce a hidden parameter τ to represent the friction at the ends of the cylinders. Then $\tau > 0$ in the experiment, but $\tau = 0$ at the organising centre, and so Figure 13 will turn out to be a section of a higher dimensional catastrophe not through the organising centre. This explains why it is stable and non-local (see 3.4). When $\tau > 0$ the fluid velocity at the ends of the cylinder has to vanish due to friction, but when $\tau = 0$ it need not vanish. This has the great advantage of allowing Couette flow ξ to become a valid solution of the equations, and so Schaeffer was then able to follow Taylor's trick of making it the organising centre.

Let R_2 be the critical value of R at which ξ becomes unstable with respect to the Taylor 2-cell flow. Taylor [71] showed that the 2-cell flow is represented by a perturbation $\xi + \eta y$, where η is a velocity field orthogonal to ξ, and y is a real variable obeying a pitchfork bifurcation with organising centre $(R_2, 0)$, as shown in Figure 15. If $R < R_2$ then ξ is an attractor, given

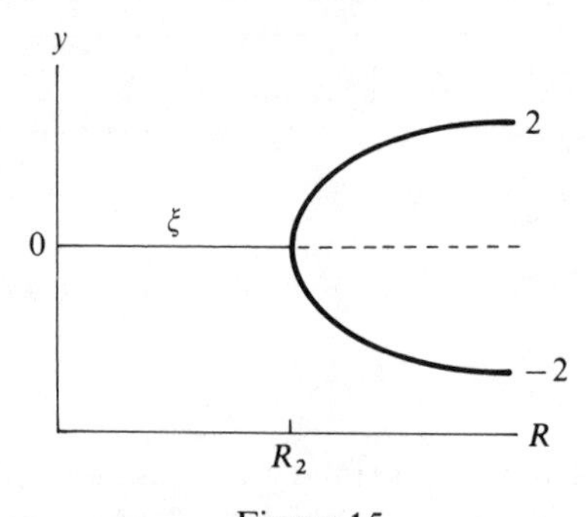

Figure 15

by $y = 0$. If $R > R_2$ then the upper branch in Figure 15 is an attractor, representing the 2-cell flow shown in Figure 11, and the lower branch is another attractor representing the similar flow spiralling in the opposite direction, with both ends spiralling inwards rather than outwards (which is equally valid when $\tau = 0$ due to the absence of any frictional bias). Similarly, let R_4 be the critical value at which ξ becomes unstable with respect to the 4-cell flow, and let $\xi + \zeta z$ be the corresponding perturbation of ξ, where ζ is a velocity field orthogonal to ξ, and z a real variable obeying another pitchfork with organising centre $(R_4, 0)$.

Now consider the double perturbation $\xi + \eta y + \zeta z$. If $R_2 \neq R_4$ the two pitchforks have different organising centres, and so the next trick is to make $R_2 = R_4$, so as to bring them together into a double-pitchfork (see Figure 5). This trick is possible because R_2 and R_4 are functions of the aspect ratio ρ,

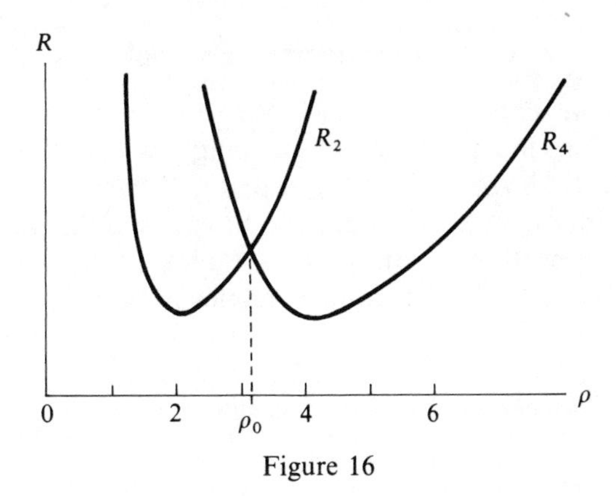

Figure 16

as shown in Figure 16. R_2 has a minimum near 2, R_4 has a minimum near 4, and their graphs cut at ρ_0, say, near 3. Putting $\rho = \rho_0$ makes $R_2 = R_4$.

In the classical language this makes the problem into a double eigenvalue problem. In the language of singularity theory: unfold the double-pitchfork into the double-cusp catastrophe, restrict to the maximal symmetric section, perturb in the direction of the asymmetric parameter τ, and take the primary branch M (see 1.4). Now ρ is the other symmetric parameter besides R, and so M will be a connected surface in the 4-dimensional (R, ρ, y, z)-space. Finally, let $v = z - y$. Then v is a suitable variable (in the sense of 5.2) and so the map $(y, z) \mapsto v$ induces an embedding of M in the 3-dimensional (R, ρ, v)-space, recovering Figure 13. Of course, v is not the same variable as we had before, but the surface is equivalent.

The reason for restricting to the maximal symmetric section is that, when $\tau = 0$, if $\xi + \eta y + \zeta z$ is an attractor, then $\xi \pm y\eta \pm \zeta z$ will also be attractors; therefore the equilibrium set is symmetric with respect to change of sign of y and z. We need to explain, however, why we have ignored the 3-cell modes, which are equally valid when $\tau = 0$, and which bifurcate off ξ at a lower Reynolds number than $R_2 = R_4$ when $\rho = \rho_0$. They can be ignored because, when $\tau > 0$, the frictional bias towards inwards spiralling at both ends has the effect of disconnecting both 3-cell modes from the primary branch M, so that they become secondary modes that will not be observed without special initial conditions. Similarly the lower branches of both the 2-cell and 4-cell pitchforks are disconnected from M, leaving in M only the two modes y, $z > 0$ illustrated in Figure 11.

5.5. Summary. Schaeffer found a hidden organising centre at which the Navier–Stokes equations could be analysed to provide a theoretical explanation for Benjamin's experimental data. It would be even more interesting if his techniques could be extended to non-elementary theory to embrace the periodic, quasi-periodic, and turbulent motion at higher Reynolds numbers [12, 13, 67]. This is the theme of the rest of the paper.

6. Non-elementary Examples

In this section we illustrate the difference between bifurcation and catastrophe in non-elementary theory by describing some examples analogous to those in Section 1. The main difference is that here the flows contain periodic attractors, or limit cycles. We define the Bowen–Ruelle measure [9] on an attractor and show that it can vary continuously even when there is a catastrophic Ω-explosion. For more examples of bifurcation see [1, 32, 36, 68, 69].

6.1. The Hopf Bifurcation [25]. This has canonical equations

$$\dot{\theta} = 1$$

$$\dot{r} = br - r^3$$

where (r, θ) are polar coordinates for the space $\mathbb{R}^2$, and b is a parameter. When $b \leq 0$ the origin O is an attractor; when $b > 0$ it becomes a repellor, and a new periodic orbit α appears at $r = \sqrt{b}$ (see Figure 17). Therefore the nonwandering set is

$$\Omega_b = \begin{cases} O, & b \leq 0 \\ O \cup \alpha, & b > 0. \end{cases}$$

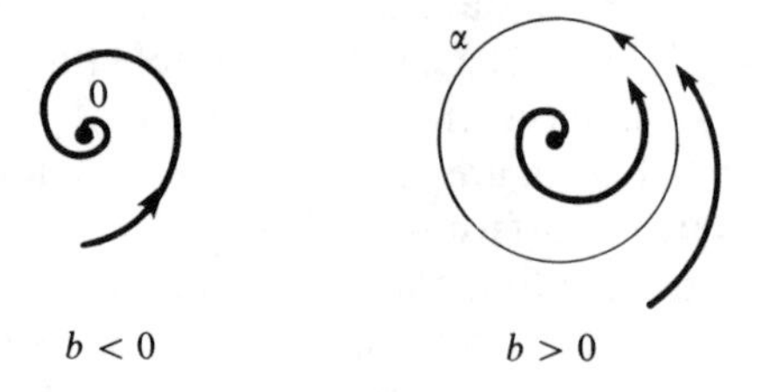

Figure 17

By Hopf's theorem [25, 32] this example is stable and local, with organising centre at the origin, $r = b = 0$. Since Ω_b depends continuously on b, the point $b = 0$ is a bifurcation point.

6.2. An Ω-Explosion. Let the state-space be the unit circle α, with coordinate θ, $0 \leq \theta < 2\pi$, and let a be a parameter. Consider the equation

$$\dot{\theta} = a - \cos\theta.$$

When $-1 < a < 1$ there is a repellor S at $\theta = \cos^{-1} a$, and an attractor A at $\theta = -\cos^{-1} a$. When $a = 1$ these two fixed points coalesce into a single fixed point N at $\theta = 0$, and when $a > 1$ they disappear (see Figure 18). Therefore the nonwandering set is

$$\Omega_a = \begin{cases} A \cup S, & -1 < a < 1 \\ N, & a = 1 \\ \alpha, & a > 1. \end{cases}$$

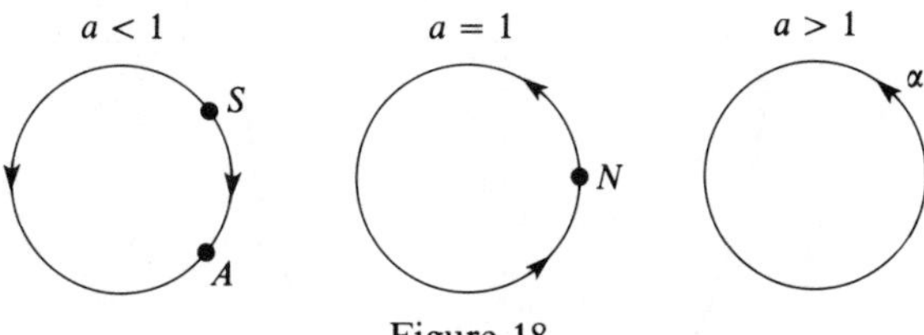

Figure 18

Although locally there is a fold catastrophe at N, globally there is a Ω-explosion from the point N to the whole circle α. Since Ω_a is discontinuous at $a = 1$ the latter is a catastrophe point. Therefore qualitatively we may call this example a catastrophe, but quantitatively there are measure-theoretic arguments for calling it a bifurcation, as we shall see in 6.6 below.

Since there is no organising centre, this example is stable and non-local. However, we can find a hidden organising centre, as follows. First embed the unit circle in $\mathbb{R}^2$ and extend the equation to:

$$\dot{\theta} = a - r \cos \theta$$
$$\dot{r} = r - r^3.$$

This preserves the attractor A, converts the previous repellor S into a saddle, and introduces a repellor R at the origin. When $a = 1$ the attractor coalesces with the saddle at the saddlenode N, and when $a > 1$ they disappear leaving only the attractor α and repellor R (see Figure 19). The extended system is

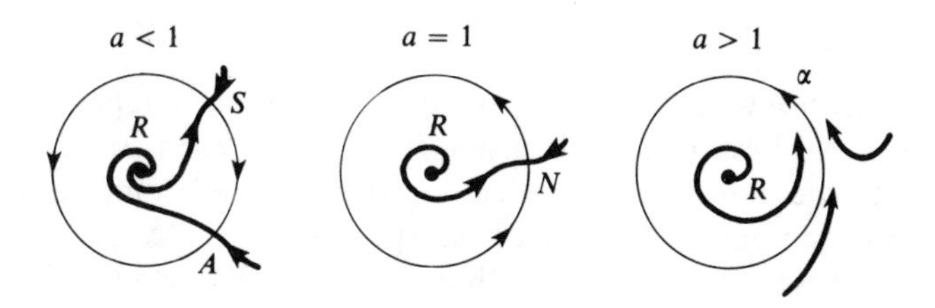

Figure 19

again stable and non-local, but we can localise it by introducing another parameter b:

$$\dot{\theta} = a - r \cos \theta$$
$$\dot{r} = br - r^3$$

This 2-parameter system is local, with organising centre at the origin; therefore the latter is a hidden organising centre for the Ω-explosion, which is given by the section $b = 1$. The bifurcation set in the parameter space consists of the line $b = 0$ of Hopf bifurcations, and the parabola $a^2 = b$ of Ω-explosions, as shown in Figure 20.

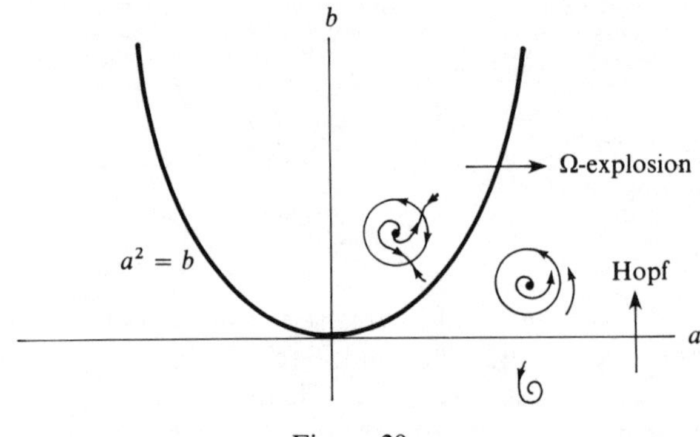

Figure 20

6.3. A Saddle Connection Catastrophe. Let the state space be $\mathbb{R}^2$, and let a be a parameter. The saddle connection is illustrated in Figure 21, and we give equations for it below, but first let us describe it qualitatively.

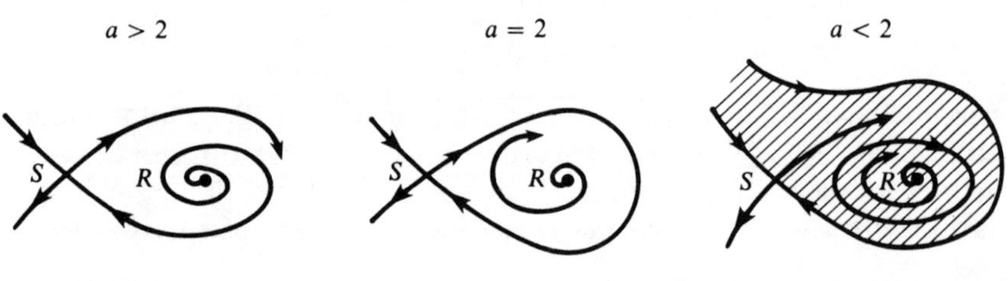

Figure 21

When $a > 2$ the nonwandering set consists of two points, a repellor R and a saddle S, and the inset of S curls inside the outset. When $a = 2$ the inset coalesces with the outset to form a saddle connection or homoclinic orbit, $\varkappa$. As a result the nonwandering set has exploded to $R \cup S \cup \varkappa$ ($\varkappa$ is nonwandering because of the orbits spiralling out from R). When $a < 2$ the inset has crossed the outset creating a new periodic attractor, α. This is a catastrophe both qualitatively and quantitatively, qualitatively because Ω_a is discontinuous at $a = 2$, and quantitatively because there is a measure-theoretic discontinuity as the new attractor captures its basin of attraction (shown shaded in Figure 21).

This example is stable and non-local, but it has a hidden organising centre at the origin of the following local 2-parameter system:

$$\dot{x} = \frac{\partial H}{\partial y}$$

$$\dot{y} = \frac{\partial H}{\partial x} + y(a - H), \quad \text{where } H = x^3 - 3bx + y^2.$$

The reader will recognise this as a parametrised damped Hamiltonian

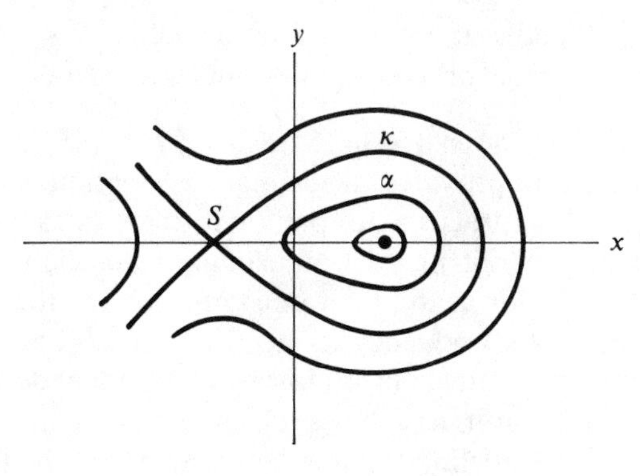

Figure 22

system [1], with Hamiltonian H and a damping that drives it to the energy level $H = a$. Figure 21 is the section given by $b = 1$. When $b = 1$ the level curves of H are shown in Figure 22; there is a minimum at $(1, 0)$ where $H = -2$, and a saddle at $(-1, 0)$ where $H = 2$. When $a \leq -2$ the minimum is an attractor of the flow, and when $a > -2$ it is a repellor; therefore when $a = 2$ there is a Hopf bifurcation at the minimum. When $-2 < a < 2$ the energy level $H = a$ contains a compact component α which is the periodic attractor shown in Figure 21. When $a = 2$ the energy level $H = 2$ contains the homoclinic orbit $\varkappa$. When $a > 2$ the energy level $H = a$ no longer contains a compact component and so the attractor disappears. Therefore $a = 2$ gives the saddle connection catastrophe, as illustrated in Figure 21.

When $b > 0$ the fixed points occur at $(\pm\sqrt{b}, 0)$ and the bifurcations occur at $a = \pm 2b\sqrt{b}$; therefore as $a, b \to 0$ the qualitative picture shrinks into the organising centre. When $a \neq 0$ and $b = 0$ the two fixed points coalesce in a fold catastrophe or saddlenode. Therefore the bifurcation set in the parameter space consists of the a-axis and the cusp $a^2 = 4b^3$, as shown in Figure 23. The negative a-axis comprises attractor-saddlenodes and the pos-

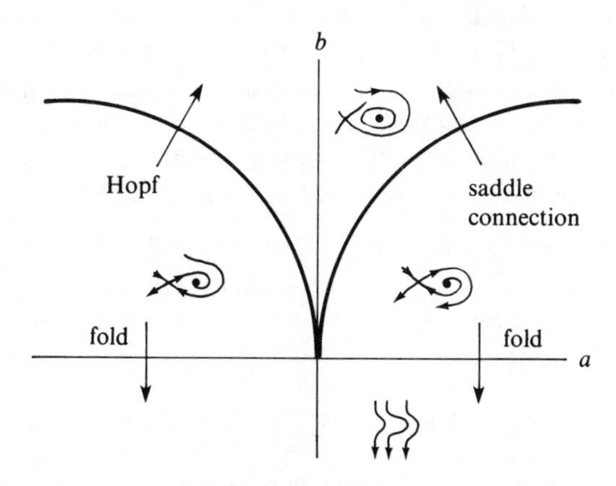

Figure 23

itive a-axis repellor-saddle-nodes; the left branch of the cusp comprises Hopf bifurcations, and the right branch saddle-connection catastrophes.

6.4. Definition of the Bowen–Ruelle Measure [9, 52, 57]. In elementary theory the phase portrait provides an adequate description of the asymptotic behaviour because the attractors are points. In the non-elementary theory, however, the phase portrait may be inadequate from the point of view of measurement because, for example, it does not provide the frequency of a periodic attractor, nor the frequency spectrum of a strange attractor. What is needed in addition to the attractor is a measure on it that describes the time spent in different parts of it, as follows. Given a flow ϕ on X, and a probability measure m on X, define the time average of m to be the measure

$$\mu = \lim_{T \to \infty} \frac{1}{T} \int_0^T \phi^t m \, dt.$$

We call μ the *Bowen–Ruelle measure.* Roughly speaking m represents the initial conditions and μ the asymptotic behaviour. For instance if the initial position is x choose m to be the Dirac measure $m = \delta_x$ with support x. If the initial position is uncertain represent this uncertainty by a suitable continuous probability measure m on X. Now for some examples of μ.

(i) If A is a point attractor and the support of m is contained in the basin of attraction of A then $\mu = \delta_A$.
(ii) If α is a periodic attractor and the support of m is contained in the basin of α then μ has support α, and at each point of α the density of μ is inversely proportional to the speed. Therefore in both these examples the Bowen–Ruelle measure exists, and is invariant, ergodic, and independent of m.
(iii) If A, B are two point attractors, and the support of m is contained in the union of their basins, and m_A, m_B are the measures of their basins, then $\mu = m_A \delta_A + m_B \delta_B$.

6.5. Parametrised Measures. Given a parametrised system, and a continuous* probability measure m, there is a time average μ_c for each parameter point c, and so we can ask the question whether or not μ_c depends continuously on c. At regular points and bifurcation points μ_c is continuous, and so it is a question of dividing the catastrophe points into those where the measure is continuous and those where it is not, as follows.

(i) At the catastrophic jump in 1.2 the measure is discontinuous because of the sudden disappearance of a basin of attraction.
(ii) At the cusp catastrophe in 1.3 the measure is continuous, but is a limit point of discontinuities.

* For parametrised systems it is better to have m continuous in order to avoid the artificial discontinuities in μ_c that arise from discontinuities in m. For example if $m = \delta_x$ and S is the separatrix between two basins, then any change in c that moves S across x will cause a jump in μ_c.

(iii) At the saddle connection catastrophe in 6.3 the measure is discontinuous because of the sudden capture of a basin of attraction. To be precise, μ_a is discontinuous as $a \searrow 2$.

(vi) At the Ω-explosion in 6.2 the measure is, surprisingly, continuous. This is the most interesting example because Ω_a is discontinuous at $a = 1$. Qualitatively it is a catastrophe, but quantitatively it behaves like a bifurcation. In an experiment one would expect to observe qualitative change, but at the same time expect quantitative time-average measurements to vary continuously. A similar phenomenon is sometimes observed at the onset of turbulence, and in the next section we shall suggest a generalisation of this example as a possible model for turbulence. Meanwhile since this is the simplest example of the phenomenon it is worth giving the proof

6.6. Lemma. *At the Ω-explosion in 6.2 the Bowen–Ruelle measure is continuous.*

PROOF. If $a = 1$ then, for any m, $\mu_a = \delta_N$ the Dirac measure at N. If $a < 1$ then, for any $m \neq \delta_R$, $\mu_a = \delta_A$, and so μ is continuous as $a \nearrow 1$. If $a > 1$ then μ_a is distributed over α with density $\sim (a - \cos\theta)^{-1}$, and so to prove $\mu_a \to \delta_N$ we have to show that the flow lingers longer and longer in the neighbourhood of N as $a \searrow 1$. More precisely, it suffices to show:

$$\forall \varepsilon > 0, \quad \exists \eta > 0, \quad \forall a,\ 1 < a < 1 + \eta \Rightarrow \mu_a[-\varepsilon, \varepsilon] > 1 - \varepsilon.$$

Let

$$A = \int_{-\varepsilon}^{\varepsilon} \frac{d\theta}{a - \cos\theta}, \qquad B = \int_{\varepsilon}^{2\pi - \varepsilon} \frac{d\theta}{a - \cos\theta}.$$

It suffices to show $B < \varepsilon A$, for then

$$\mu_a[-\varepsilon, \varepsilon] = \frac{A}{A + B} = 1 - \frac{B}{A + B} > 1 - \frac{B}{A} > 1 - \varepsilon.$$

Let $K = \int_{-\varepsilon}^{\varepsilon} (d\theta/1 - \cos\theta)$. If $a > 1$ then $B < K$. If $a < 1 + \eta$ then

$$A > \int_{-\varepsilon}^{\varepsilon} \frac{d\theta}{(1 + \eta) - (1 - \theta^2/2)} = \frac{4}{\sqrt{2\eta}} \tan^{-1} \frac{\varepsilon}{\sqrt{2\eta}} \xrightarrow[\eta \to 0]{} \infty.$$

If η is sufficiently small then $A > K/\varepsilon$. Therefore, $B < K < \varepsilon A$, as required. This completes the proof that μ_a is continuous at $a = 1$. □

7. Strange Attractors

In this section we define and construct some strange attractors. We give both a topological definition 7.1 and a measure theoretic definition 7.8. The strange attractors that are best understood are those that satisfy axiom A

(see 7.6), but models of turbulence are more likely to need non-axiom A attractors (see 7.9). In particular the onset of turbulence would be modelled by a strange bifurcation, and in 7.11 we describe an example in which a periodic attractor runs into a strange saddle causing an Ω-explosion into a strange attractor. For further examples of strange attractors see [4, 7, 8, 20, 23, 24, 28, 29, 34, 38, 42, 45, 50, 55, 60, 61, 64, 66, 74, 75].

7.1. Topological Definition of Attractor. An *attractor* Λ of a flow ϕ on X is a subset that is attracting and indecomposable. Here *attracting* means $\exists$ a closed positively-invariant* neighbourhood N of Λ such that $\bigcap_{t>0} \phi^t N = \Lambda$. *Indecomposable* means $\exists$ a point in Λ whose ω-limit* is Λ.

It follows that Λ is a closed invariant subset of the nonwandering set Ω. One can also deduce that Λ is minimal in the sense that if one attractor contains another, or if two meet, then they are equal. Define the *basin of attraction* of Λ to be the set of points whose ω-limit is contained in Λ; it follows that the basin is an open invariant set containing N. Call Λ *stable* if perturbations of ϕ have an equivalent attractor nearby.

7.2. Examples. There are three types of *familiar* attractors, and all the rest are called *strange* attractors (because they have only really been studied in the last 20 years or so). The familiar attractors are:

(i) *Point Attractor.* A point attractor is stable if hyperbolic; here hyperbolic means the eigenvalues have non-zero (and hence negative) real part.
(ii) *Periodic Attractor.* Here the attractor is diffeomorphic to the circle $T = \mathbb{R}/\mathbb{Z}$, and is stable if hyperbolic.
(iii) *Quasi-periodic Attractor.* Here the attractor is diffeomorphic to an n-torus, $T^n = \mathbb{R}^n/\mathbb{Z}^n$, $n \geq 2$, and the flow is equivalent to an irrational flow, given by $\phi^t x = x + \omega t$, where $\omega = (\omega_1, \ldots, \omega_n)$ is constant and the ω_i are irrationally related. Therefore all orbits are dense. A quasi-periodic attractor is unstable because any rational perturbation has all orbits periodic. Moreover, if we perturb further by adding a suitable small transverse field then we obtain a stable flow with a single periodic attractor L, whose basin is dense in T^n. Such an L is called a *lock-on* because it locks all the phases of the original n oscillators together; the collapse of the attractor $T^n \to L$ is called an *Ω-implosion*.

Before we go on to construct a strange attractor we make a remark about definition 7.1 above.

Remark. There is no universally accepted definition of attractor yet because the subject is still developing. Most definitions are either too strong because they exclude important examples, or too weak because they include un-

* *Positively-invariant* means $\phi^t N \subset N$, $\forall t > 0$. The *ω-limit* of x means $\bigcap_t$ (closure $\bigcup_{s>t} \phi^s x$).

wanted examples. For instance, that in [72, page 39] is slightly too strong because it inadvertently excludes strange attractors; that in [64, page 786] is only concerned with axiom A and excludes non-axiom A attractors; that in [1, page 517] is slightly too weak because it lacks minimality, and includes, for example, all closed invariant sets of the equation $\dot{x} = -x$ on $\mathbb{R}^n$. The definition above may need to be weakened when studying strange bifurcations. Also some authors prefer a more measure-theoretic definition, like 7.8 below, because it is more closely related to the measurements made in experiments. On the other hand, it is important to keep both the topological and the measure-theoretic viewpoints in mind because the former provides an overall grasp, while the latter relates to data.

We shall now construct a strange attractor by suspending a diffeomorphism. This is the simplest method of constructing one, and indeed the study of diffeomorphisms was pioneered by Smale [64] in order to gain insight into differential equations. The most elegant attractors are obtained from Anosov diffeomorphisms [4, 30, 64], but these are exceptional because they are manifolds and do not display the Cantor set structure that is typical of most strange attractors; so instead we shall describe some examples that are more likely to be of use in modelling turbulence.

7.3. Definition of Suspension. Given a smooth embedding $f\colon M \to M$ we suspend this to a flow Σf on a manifold $\Sigma_f M$ one dimension higher, as follows. Let $F\colon \mathbb{R} \times M \to \mathbb{R} \times M$ be the map $F(s, x) = (s - 1, fx)$ and define $\Sigma_f M$ to be the quotient manifold $(\mathbb{R} \times M)/F$. Let ϕ be the flow on $\mathbb{R} \times M$ given by $\phi^t(s, x) = (s + \omega t, x)$, where $\omega > 0$. We call ω the *frequency* of the suspension. Since ϕ commutes with F it induces a flow on $\Sigma_f M$, which defines Σ_f.

If f is a diffeomorphism there is a simpler way of constructing $\Sigma_f M$ by glueing the ends of $I \times M$ together with f, so that $(1, x) = (0, fx)$ (see Figure 24).

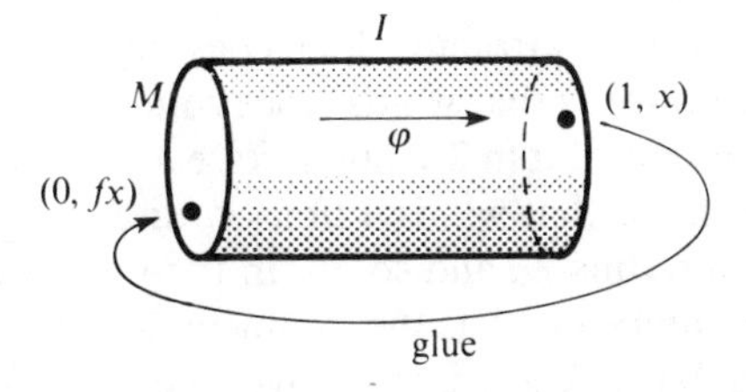

Figure 24

The fixed points and (discrete) periodic orbits of f suspend into the (continuous) periodic orbits of Σf. The attractors of f suspend into the attractors of Σf. Here an attractor of an embedding is defined in the same way, word for word, as an attractor of a flow (with the understanding in 7.1 that t lies in $\mathbb{Z}$ rather than $\mathbb{R}$).

7.4. Example of a Strange Attractor [61, 64]. Let M be the 3-dimensional solid torus

$$M = \{(w, z);\ w, z \in \mathbb{C},\ |w| = 1,\ |z| \leq 1\}.$$

Define the embedding $f: M \to M$ by

$$f(w, z) = \left\{w^2, \frac{w}{2} + \frac{z}{4}\right\}.$$

Let

$$\Lambda = \bigcap_{n > 0} f^n M$$

(see Figure 25). We shall show that Λ is a strange attractor of f. Therefore if we suspend f to give a flow Σ_f on the 4-dimensional manifold $\Sigma_f M$, then $\Sigma\Lambda$ will be a strange attractor of the flow.

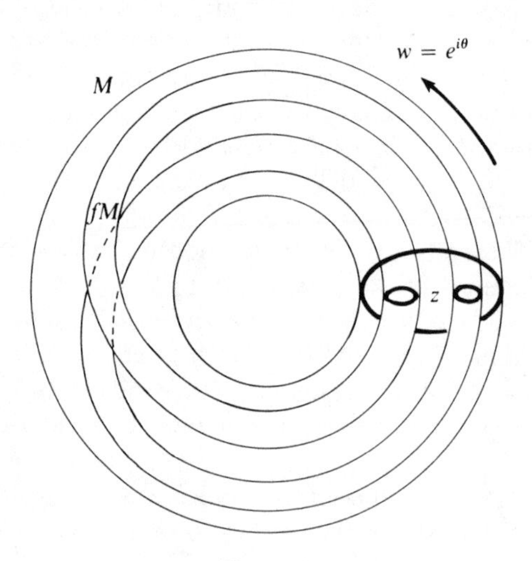

Figure 25

Before proving Λ is an attractor, let us describe its geometry. The image fM of f is a long thin torus that winds twice round inside M, and meets each transverse disk w = constant in 2 small disks, each of radius $\frac{1}{4}$. Similarly f^2M is a longer thinner torus that winds 4 times round, and meets each disk in 4 small disks, each of radius $\frac{1}{16}$, and so on. In the limit Λ meets each disk in a Cantor set. Therefore locally Λ is the 1-dimensional product of an arc and a Cantor set, and globally Λ is a bundle over the circle with fibre the Cantor set (like a solenoid). Consequently the suspension $\Sigma\Lambda$ is locally the 2-dimensional product of a surface and a Cantor set.

7.5. Lemma. *Λ is a stable attractor of f.*

Proof. By construction Λ is attracting, and so it suffices to prove it indecomposable. Following Parry we prove Λ indecomposable by showing $f \mid \Lambda$

conjugate to a shift automorphism, as follows. Let $2^{\mathbb{Z}}$ denote the space of all doubly-infinite sequences of 0's and 1's, with the compact-open topology. Let σ be the left-shift on $2^{\mathbb{Z}}$ given by $(\sigma y)n = y(n+1)$, $y \in 2^{\mathbb{Z}}$, $n \in \mathbb{Z}$. Given

$$x = (e^{i\theta}, z) \in \Lambda \quad \text{let } h_0 x = \begin{cases} 0, & 0 \leq \theta < \pi \\ 1, & \pi \leq \theta < 2\pi. \end{cases}$$

Applying h_0 to the orbit $[\ldots, f^{-1}x, x, fx, f^2x, \ldots\}$ of x determines a map $h: \Lambda \to 2^{\mathbb{Z}}$, which is not continuous, but which makes the following diagram commutative:

$$\begin{array}{ccc} \Lambda & \xrightarrow{f} & \Lambda \\ {\scriptstyle h}\downarrow & & \downarrow{\scriptstyle h} \\ 2^{\mathbb{Z}} & \xrightarrow{\sigma} & 2^{\mathbb{Z}} \end{array}$$

Now $h\Lambda$ consists of all sequences except those ending in an infinite string of 1's. Therefore, as in decimals, let us identify a sequence ending 0111... with the sequence having the same beginning but ending 1000..., and identify the sequence of all 1's with that of all 0's. Let Λ^* be the identification space, and $v: 2^{\mathbb{Z}} \to \Lambda^*$ the identification map. It is straightforward to verify that although h is not continuous the composition $vh: \Lambda \to \Lambda^*$ is in fact a homeomorphism. Therefore $f|\Lambda$ is conjugate to the left-shift on Λ^*. The periodic orbits in $h\Lambda$ are dense in $2^{\mathbb{Z}}$, and hence it is easy to construct a point in $h\Lambda$ with ω-limit $2^{\mathbb{Z}}$. Therefore periodic orbits are dense in Λ^*, and Λ^* is indecomposable. Therefore the same is true for Λ, and so Λ is an attractor. For the proof of stability the reader is referred to [39, 44, 65]. □

7.6. Definition of Axiom *A* Attractor. Following Smale [64] we say an attractor Λ of a flow on X satisfies *axiom A* if periodic orbits are dense in Λ and Λ has a hyperbolic structure; here a *hyperbolic structure* means that at each point Λ there is

$$\begin{cases} 1 \text{ dimension of flow} \\ e \text{ dimensions of expansion} \\ n-1-e \text{ dimensions of contraction} \quad (n = \dim X), \end{cases}$$

and that this decomposition is continuous (see [30, 64] for details). For instance example 7.4 above satisfies axiom A: the periodic orbits are dense by the lemma, and it is hyperbolic because there is 1 dimension of flow in the suspension direction, 1 dimension of expansion in the w-direction, and 2 dimensions of contraction in the z-direction. Locally an axiom A attractor is always like a manifold in the flow and expansion directions, but may be like a submanifold, or a Cantor set, of a product of the two, in the contraction direction.

In a familiar attractor there is no expansion and so $e = 0$, but in a strange attractor these must always be an expanding direction, and so $e \geq 1$. This expanding quality is an extremely important property of strange attractors because points that are close together are torn apart exponentially. Therefore it implies *sensitive dependence on initial condition* [54], and gives an unpredictable and chaotic appearance to the motion [28, 50, 55, 60]. At the same time the advantage of axiom A is that it is a sufficient condition for the attractor to be stable [39, 65]. Hence we have that paradoxical combination of chaos and stability which is so noticeable in turbulence. Axiom A attractors also have nice measure theoretical properties, as follows. Recall the Definition 6.4 of the Bowen–Ruelle measure.

7.7. Theorem (Bowen–Ruelle [9, 52]). *If Λ is an axiom A attractor with basic B then for any continuous probability measure m with support in B the Bowen–Ruelle measure μ exists, has support Λ, and is invariant, ergodic, and independent of m.*

Remark. If an attractor Λ is strange there must be some exceptional points in its basin B whose ω-limit is not the whole of Λ; for instance each periodic orbit inside Λ must be the ω-limit of some exceptional points. If x is exceptional the time average of δ_x cannot equal the Bowen–Ruelle measure μ on Λ—that is why we required m to be continuous in Theorem 7.7. However, if Λ satisfies axiom A these exceptional points only have Lebesgue measure zero. Therefore the complementary set B_0 of generic points has Lebesgue measure 1 in B, and $\forall x \in B_0$, $\mu =$ the time average of δ_x. In other words, most initial conditions lead to μ.

Now suppose $g\colon B \to \mathbb{R}$ is a continuous function representing some experimental measurement. Define the time average $\bar{g}$ of g to be

$$\bar{g}x = \lim_{T \to \infty} \frac{1}{T} \int_0^T g(\phi^t x)\, dt.$$

Then Theorem 7.7 implies that $\bar{g}$ is constant on B_0 and

$$\bar{g}x = \int_\Lambda g\, d\mu, \qquad \forall x \in B_0.$$

The significance of this result is that in spite of sensitive dependence on initial condition, the time-average measurements are not sensitive. Although the flow may look chaotic the measurements are meaningful and repeatable. Rand [47] suggests that since this property of attractors is the most important one from the experimental point of view it ought to be embodied in the definition, in order to distinguish those attractors that might be useful for modelling. He therefore proposes that the following definition might be a fruitful extension of axiom A.

7.8. Measure-Theoretic Definition of Attractor. Given a flow ϕ on X and an open set $B \subset X$ we say *there is an attractor in B* if the time average of any function B is almost constant. Here *almost constant* means $\exists$ a subset B_0, of

Lebesgue measure 1 in B, such that, $\forall$ continuous $g: B \to \mathbb{R}$, $\bar{g}$ is constant on B_0. It follows that the Bowen–Ruelle measure μ exists, and, $\forall g$, $\bar{g} = \int g \, d\mu$. The *attractor* Λ is defined to be the support of μ.

7.9. The Henon Attractor [23]. Non-axiom A attractors are relatively unexplored but are the subject of much current research. The best known examples are the Lorenz attractor [20, 28, 45, 53, 75] and the Henon attractor, although neither of these has been proved to satisfy either of the above definitions yet.

The Henon attractor is a closed invariant set Λ of the diffeomorphism $f: \mathbb{R}^2 \to \mathbb{R}^2$ given by $f(x, y) = (y + 1 - ax^2, bx)$, where $a = 1.4$, $b = 0.3$. Henon [23] proved that Λ is attracting, and used a computer to show it looks like Figure 26. However Λ has not yet been proved indecomposable,

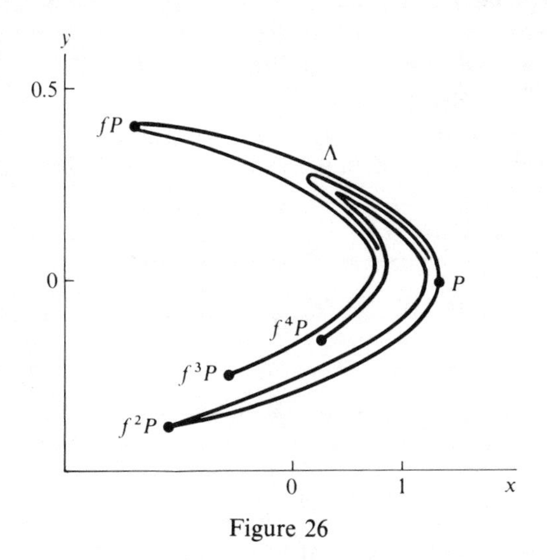

Figure 26

although numerical results suggest that it is. The computer pictures look as if Λ is locally the product of an arc and a Cantor set, but this is unlikely to be true for the following reason. If $P \in \Lambda$ is a point on the x-axis, then it can be seen from Figure 26 that the arc through P gets folded sharper and sharper under the iterations of f, so that, if the images of P are dense in Λ, then any open subset of Λ must contain an infinity of them and cannot therefore be a product. This phenomenon also prevents Λ from being hyperbolic. Therefore it does not satisfy axiom A and is difficult to handle with Smale theory; however, it may be possible to show it is a measure-theoretic attractor in the sense of 7.8 by using characteristic exponents and the more general Pesin theory [41, 56]. For some values of the parameters (a, b) computer studies suggest that Λ breaks up into a (possible infinite [37]) number of periodic attractors. Even if this were to happen for a dense set of parameters nevertheless it may be possible to estimate a bound for the variation of the associated Bowen–Ruelle measure μ, which is really what is needed from the experimental point of view.

Meanwhile Lozi [29] has studied the piecewise-linear analogue of the Henon attractor, given by replacing x^2 by $|x|$ in the definition of f, and Misiurewicz [34] has shown it indecomposable for an open set V of parameters. Although the Lozi attractor is not smooth it does satisfy Definition 7.1, and probably the associated Bowen–Ruelle measure is continuous in V.

The map f in the Henon attractor is orientation reversing, but an orientation preserving analogue is obtained by changing the sign of b, and then the suspension Σf gives a 2-dimensional strange attractor in a 3-dimensional solid torus, which is much closer to the type of attractor we are looking for to model turbulence.

7.10. The Horseshoe [62, 63, 64]. If we are to model the onset of turbulence by a strange bifurcation, it is necessary to study strange saddles as well as strange attractors. The most famous strange saddle is the Smale horseshoe, which is defined as follows.

Let M be the union of a square Q and two semi-disks D, D'. Let f: $M \to M$ be a horseshoe-shaped embedding, as shown in Figure 27, such that on the

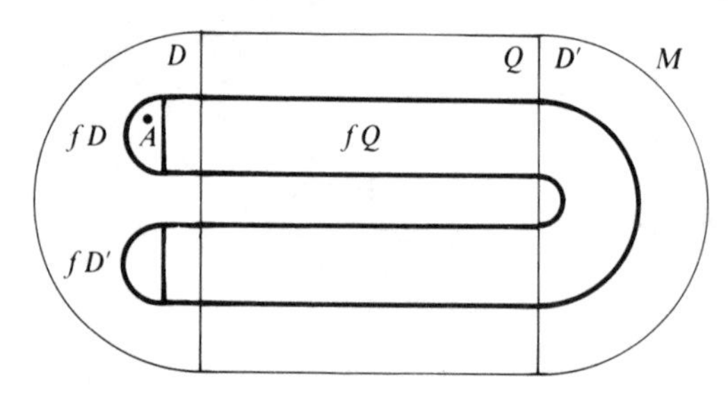

Figure 27

semi-disks f is contracting, and on $f^{-1}Q \cap Q$ it is linear and hyperbolic, expanding horizontally and contracting vertically. Then there is a unique point attractor $A \in fD$, and the rest of the nonwandering set is the strange saddle H, defined as follows.

Let $Q_0 = Q$, let $Q_{n+1} = f^{-1}Q_n \cap fQ_n$ inductively, and let $H = \bigcap_n Q_n$. Then Q_1 comprises 4 small rectangles, Q_2 comprises 16 smaller rectangles, and so on, until in the limit H is the product of two Cantor sets, and hence is itself a Cantor set (see Figure 28). H is invariant, indecomposable and densely filled with periodic points, because $f|H$ is conjugate to the left-shift on $2^{\mathbb{Z}}$ (as in Lemma 7.5). H is hyperbolic by construction, and so satisfies axiom A. Therefore H is a strange saddle, and stable [63].

In particular H contains two fixed points S, S' which are both saddle points. S is the top left corner of H and S' is near the bottom right corner. Let $\Lambda = \bigcap_{n>0} f^nM$, as shown in Figure 29. In fact Λ is the union of the sink A and the outset of H. Locally Λ is the product of an arc and a Cantor set,

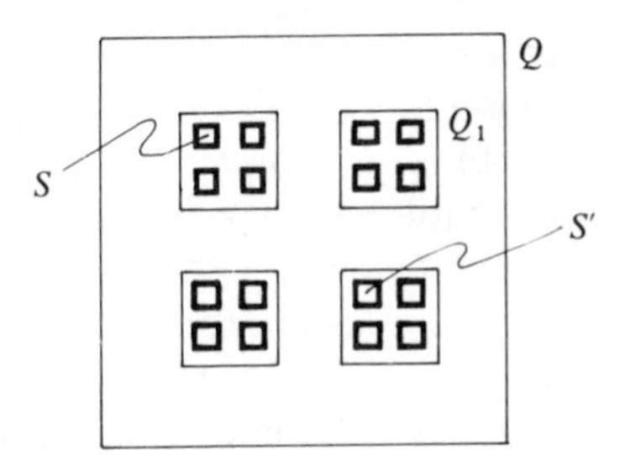

Figure 28

except at A. Notice that Λ is attracting but it is not an attractor because it is not indecomposable. In fact the points of $\Lambda - H - A$ are wandering, and wander off to A. The outsets of S, S' are contained in Λ; the left outset of S is the interval SA running straight into A, while the right outset of S and both outsets of S' wind densely over Λ.

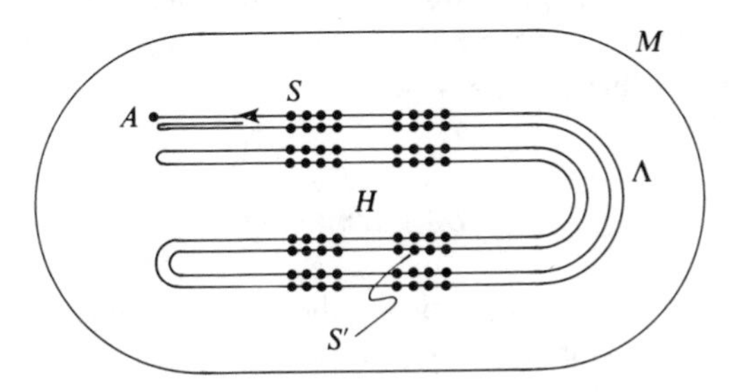

Figure 29

7.11. The Sink-Horseshoe Bifurcation. This is an example of a strange bifurcation which might be useful in modelling, but which is not yet fully understood mathematically and needs to be studied further. Introduce a parameter in the above construction of the horseshoe, and use the parameter to change f in the neighbourhood of AS so as to run the sink A into the saddle point S. Locally this is the simplest form of catastrophe, namely a saddlenode or fold catastrophe, but globally we must regard it as a sink running into the strange saddle H, because H is indecomposable. It is analogous to the Ω-explosion of Example 6.2, because the nonwandering set explodes* from $A \cup H$ into Λ. Roughly speaking after the bifurcation all the

* More precisely, let c be the parameter, $c = 0$ the bifurcation point, Ω_c the nonwandering set of f_c, $\Omega = \bigcup c \times \Omega_c$, $\bar{\Omega} = \text{closure } \Omega$, $\bar{\Omega}_c = \bar{\Omega} \cap (c \times M)$, and $\Lambda_c = \bigcap_n f_c^n M$. Then

$$\Omega_c = \bar{\Omega}_c = A_c \cup H_c, \qquad c > 0$$

$$\Omega_0 = H_0$$

$$\bar{\Omega}_0 = \Lambda_0 .$$

We know $\Omega_c \subset \Lambda_c$, $\forall c$, and $\Omega_c \neq \Lambda_c$ for some positive values of c near 0, but conjecture that $\Omega_c = \Lambda_c$ for some other positive values of c near 0.

points of Λ which used to wander off to A now re-enter again at S and go round Λ again. After the bifurcation S has disappeared and so an exponentially thin neighbourhood of the outset of S has been removed from Λ, but S' is preserved, and Λ is probably still the closure of its outset. In fact Λ resembles the Henon attractor 7.9 (with $b < 0$). For all parameter values Λ is attracting but for some values after bifurcation it cannot be an attractor, because it contains periodic sinks with very long periods and very small basins [37], and is therefore not indecomposable. However we conjecture* that the Bowen–Ruelle measure is continuous at the bifurcation point, as in Lemma 6.6.

Suppose that we now suspend the whole picture to give a parametrised flow Σ_f on the 3-dimensional open solid torus $\Sigma_f M$. Before bifurcation ΣA is a periodic attractor in the torus representing periodic motion, and after bifurcation $\Sigma\Lambda$ is (or, more precisely, resembles) a 2-dimensional strange attractor representing turbulence. At bifurcation the Ω-explosion represents the qualitative catastrophic onset of turbulence, and the μ-continuity represents the quantitative continuity of time-average measurements. We pursue this model further in 8.4 below.

8. Turbulence

Ruelle and Takens [51] first proposed the use of strange attractors to model turbulence in 1971. This intriguing idea has attracted a great deal of attention, but as yet the programme is still in its infancy: much of the mathematics is still unresolved, and the application is mostly speculative. The future mathematical programme will require:

(i) topological studies of strange bifurcations and their organising centres (extending the ideas of Sections 5 and 6);
(ii) measure theoretic studies of strange bifurcations, generalising Fourier analysis from periodic attractors to strange attractors;
(iii) quantitative analysis of the Navier–Stokes equations at the strange organising centres (as in Section 4), leading to prediction and experimental confirmation.

The central idea of the programme is to provide a conceptually simple geometric link between the complexity of the Navier–Stokes equations on the one hand and the complexity of the observed data on the other. In this section we discuss some points in the programme. For further discussions see [7, 8, 12, 13, 14, 26, 28, 35, 47, 50, 53, 54, 55, 57, 60, 66, 67, 70].

8.1. The Ruelle–Takens Model. We begin with the Landau–Lifschitz model [26] for the onset of turbulence. Let X be the space of all possible fluid

* This conjecture is the reason for calling it a bifurcation rather than a catastrophe.

velocity fields in a given region, and let E be the evolution equation on X determined by the Navier–Stokes equations and parametrised by the Reynold's number R. For $R < R_1$ suppose that E has a point attractor T^0, representing steady fluid motion. At R_1 there is a Hopf bifurcation $T^0 \to T^1$, to a periodic attractor T^1 representing periodic fluid motion. If the Reynolds number is increased to R_2 there is another Hopf bifurcation $T^1 \to T^2$, to a quasi-periodic attractor on a torus T^2 (see 7.2). At R_3 here is further Hopf bifurcation $T^2 \to T^3$, to a quasi-periodic attractor on a 3-dimensional torus T^3, and so on, giving the sequence of bifurcations

$$T^0 \to T^1 \to T^2 \to T^3 \to \cdots$$

as the Reynolds number increases, leading eventually to turbulence.

Ruelle and Takens criticised the Landau–Lifschitz model on the grounds that since T^n is unstable for $n \geq 2$ it is unlikely to be observed. They showed [38, 51] that for $n \geq 3$ there exist stable perturbations $T^n \to \Lambda$, where Λ is a strange attractor in T^n of lower dimension. Then the strange attractor will exhibit the desired sensitive dependence on initial condition [54]. Although this was a far reaching idea, the Ruelle–Takens model can itself be criticised on three grounds, as follows.

(i) Applying their own criticism and construction to the case $n = 2$ gives a lock-on $T^2 \to L$, where L is a periodic attractor (see 7.2). This modifies the Landau–Lifschitz sequence to $T^0 \to T^1 \to T^2 \to L$, so that we never actually reach the situation T^n, $n \geq 3$, where the Ruelle–Takens construction can be used to obtain a strange attractor.

(ii) The particular examples [42] of strange attractors that they used to prove the mathematical existence theorem are not particularly plausible from the hydrodynamic point of view (not that they claimed any such plausibility).

(iii) At the onset of turbulence new fluid motions appear that were not observed before turbulence, and so this suggests an Ω-explosion rather than an Ω-implosion $T^n \to \Lambda$.

8.2. Experimental Data. The most interesting data obtained so far are that of Gollub, Swinney and their collaborators [12, 13, 14, 67]. They measured frequency spectra in Bénard and Taylor flows, and inferred that the following are amongst typical routes to turbulence:

$$\text{Bénard:} \quad T^0 \to T^1 \to T^2 \to L \to \text{turbulence}$$

$$\text{Taylor:} \quad T^0 \to T^1 \to T^2 \to \text{turbulence}.$$

The Bénard experiment [14] consists of heating a rectangular cell of water from below, the parameter in this case being the Rayleigh number or temperature difference. Firstly, there is a steady motion T^0, consisting of rolls parallel to the longer side, as shown in Figure 30. Secondly, when the temperature is increased the rolls begin to oscillate to and fro with a frequency

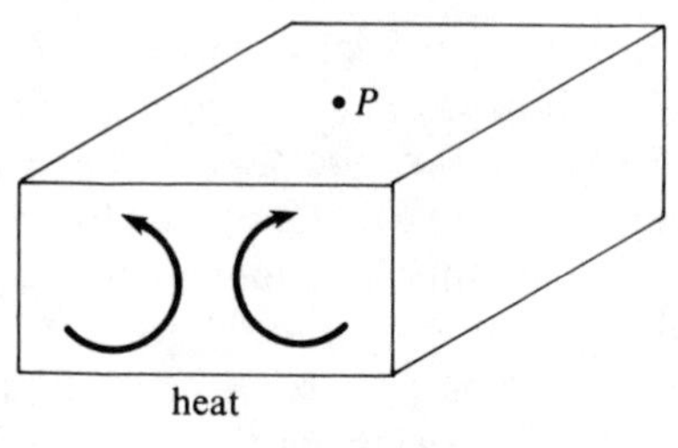

Figure 30

ω_1, say, which is represented by a bifurcation $T^0 \to T^1$ to a periodic attractor. Thirdly, the amplitude of the oscillations begins to modulate with frequency ω_2, say, which is represented by a bifurcation $T^1 \to T^2$ to a quasi-periodic attractor. Fourthly, there is a phase lock between oscillations and modulation $T^2 \to L$, both the original frequencies now being multiples of the lock-on frequency ω. Fifthly, the onset of turbulence is accompanied by a broadening of the spectral lines to bands. A typical sequence of frequency spectra is sketched in Figure 31. [See 12, 14.]

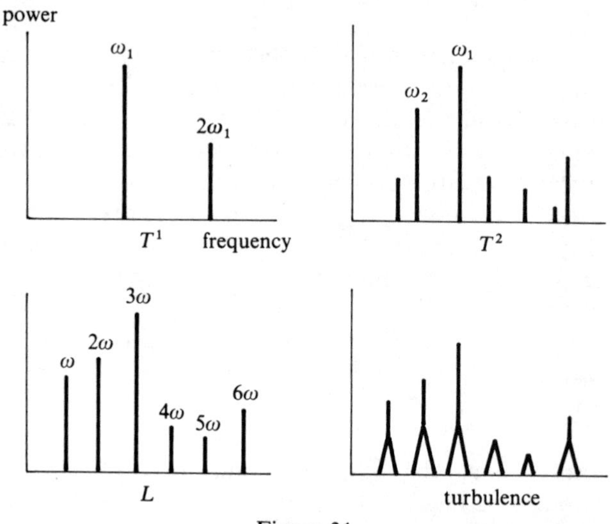

Figure 31

These spectra are obtained by monitoring some component v of the fluid velocity at a suitable point, for instance the horizontal component at P in Figure 30. The velocity is calculated from the Doppler effect on a laser beam focused at P and reflected off tiny polystyrene balls moving with the fluid. Then the spectrum is the Fourier transform of the autocorrelation function of v (see 8.3 below).

In the Couette–Taylor experiment T^0 represents the steady motion of Taylor flow described in Section 4, above. When the Reynolds number is increased the Taylor cells develop waves which rotate at (mysteriously) one third the angular velocity of the inner cylinder, which is represented by a bifurcation $T^0 \to T^1$ to a periodic attractor. Next the amplitude of the waves begins to modulate, which is represented by $T^1 \to T^2$. Finally the onset of turbulence is accompanied by a broadening of the spectral lines. There is no lock-on in this case because T^2 is stable with respect to the rotational symmetry of the apparatus [18, 46]. We now have to explain why a strange bifurcation causes a broadening of the spectral lines.

8.3. Fourier Analysis. Let ϕ be a flow on X and $v: X \to \mathbb{R}$ a measurement. If the initial condition is $x \in X$ then the autocorrelation function $\alpha: \mathbb{R} \to \mathbb{R}$ of v is defined by

$$\alpha(t) = \lim_{T\to\infty} \frac{1}{T}\int_0^T v(\phi^s x)v(\phi^{s+t}x)\,ds.$$

Assuming x is a generic initial condition in the basin of an attractor Λ with Bowen–Ruelle measure μ, then

$$\alpha(t) = \lim_{T\to\infty} \frac{1}{T}\int_0^T g^t(\phi^s x)\,ds, \quad \text{putting } g^t(x) = v(x)v(\phi^t x),$$

$$= \int_\Lambda g^t\,d\mu, \quad \text{which is independent of } x.$$

The frequency spectrum σ is the Fourier transform of α, and we now describe some examples.

(i) If $\Lambda = T^1$, a periodic attractor of frequency ω, then

$$\sigma = \sum_{n=1}^{\infty} \sigma_n \delta_{n\omega},$$

where $\delta_{n\omega}$ is the Dirac measure at $n\omega$ and σ_n the energy in the nth harmonic. Therefore the spectrum consists of lines at the multiples of ω, with heights σ_n, as in the first and third graphs of Figure 31. If we change the measurement v this merely alters the heights of the spectral lines. If ϕ depends upon a parameter like the Reynolds number, and we perturb the parameter, then this may move Λ in X and alter the fundamental frequency ω; it may also alter the speed with which ϕ flows round Λ differently in different parts of Λ, and hence change the harmonics, in other words, alter the heights of the spectral lines, but it will not change the type of σ.

(ii) If $\Lambda = T^2$, a quasi-periodic attractor with frequencies ω_1 and ω_2, then

$$\sigma = \sum_{n_1, n_2} \sigma_{n_1, n_2} \delta_{n_1\omega_1 + n_2\omega_2}$$

as in the second graph of Figure 31. Changing the measurement may alter the heights of the spectral lines, and perturbing the parameter may alter their heights and change the fundamental frequencies ω_1 and ω_2, but as long as Λ remains quasi-periodic it will not change the type of σ. This is particularly noticeable in modulated wavy Taylor flow because the symmetry makes T^2 stable [18, 46].

(iii) If the attractor is a suspension $\Sigma\Lambda$ of frequency ω, and if the measurement v depends only on the suspension coordinate, then σ will be the same as in case (i), giving the appearance of a periodic attractor. If, however, $\Sigma\Lambda$ is in fact a strange attractor, and if a perturbation of the parameter causes the speed to alter differently in different parts of $\Sigma\Lambda$ then this will uniformly broaden each spectral line into a band, as in Figure 31, because the expanding property of the strange attractor causes mixing and destroys the correlation after some time [40, 47]. For example if the change of speed caused an ε-change in the suspension frequency that depended linearly on Λ then this would destroy the correlation after time $\sim (1/\varepsilon)$, and hence broaden each spectral line into a band of width $\sim \varepsilon$. The greater the change of speed the broader the bands, until they eventually overlap and lose their identity.

8.4. Onset of Turbulence. We return to the main problem of finding bifurcations that model the onset of turbulence. Suppose that turbulence is preceded by periodic motion, represented by a periodic attractor T^1 of the evolution equation E on X. Let M be a disk of codimension 1 in X, cutting T^1 transversally at A, say. Let $f\colon M \to M$ be the Poincaré return map determined by E. Then A is a point attractor of f, and its suspension $\Sigma A = T^1$. Assume that A is stable, so that all its eigenvalues have negative real part.

In the Landau–Lifschitz model the next step $T^1 \to T^2$ is equivalent to assuming a Hopf bifurcation of f at A, which is preceded by the weakening of the attractive power of a pair of complex eigenvalues just before they cross the imaginary axis. An alternative assumption is the weakening of a single real eigenvalue. This is particularly plausible in the case when T^1 has arisen from a lock-on $T^2 \to T^1$, because then the eigenvalue of T^1 in T^2 is already weak compared with the rest.

When a real eigenvalue crosses the imaginary axis from negative to positive it produces a pitchfork bifurcation of f, as in Figure 1. (Here the equilibrium set for the parametrised embedding is equivalent to that for the parametrised flow in 1.1.) Now the pitchfork is unstable, so we look for a bifurcation arising from a stable perturbation of the pitchfork. In elementary theory the only stable perturbation of the pitchfork that contains any bifurcations or catastrophes on its primary branch is that shown in the middle picture of Figure 14; here the attractor A runs into a saddle causing a catastrophic jump to another attractor. In non-elementary theory we must allow for the possibility that the saddle might be strange, in which case we should obtain an Ω-explosion similar to the sink-horseshoe bifurcation 7.11. Indeed this example is not implausible as we now explain.

Given a weak real eigenvalue, the Poincaré return map can be written as

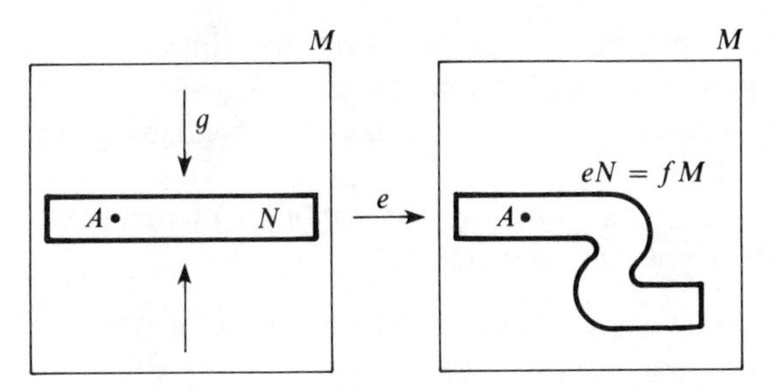

Figure 32

a composition $f = eg$, where g is a strong contraction onto a tubular neighbourhood N of the corresponding eigenvector, and e is an embedding of N in M. Locally near A the embedding maps N along the eigenvector, but globally e may bend N, as in Figure 32. Indeed the return map of a Hopf bifurcation has a similar S-bend, except that in the Hopf case A lies in the middle of the S-bend. Therefore Figure 32 could be regarded as a perturbation of the Landau–Lifschitz model, and it would be interesting to find a common organising centre.

It is possible to obtain Figure 32 by extending Figure 27, and in this case the nonwandering set of f will be the attractor A and a horseshoe as in 7.10. Then the required perturbation of the pitchfork will be none other than the sink-horseshoe bifurcation $A \to \Lambda$ described in 7.11. The suspension $\Sigma A \to \Sigma\Lambda$ is an Ω-explosion from the periodic attractor T^1 to the strange attractor $\Sigma\Lambda$ modelling the onset of turbulence.

Now in an experiment the experimenter will naturally set up his apparatus so as to best detect T^1 and measure its frequency before bifurcation. Therefore immediately after bifurcation the apparatus will automatically pick up the suspension coordinate of $\Sigma\Lambda$, and measure its frequency. Hence, assuming the Bowen–Ruelle measure continuous at bifurcation, the spectral lines will exhibit no discontinuity at the Ω-explosion. Beyond bifurcation the speed of $\Sigma\Lambda$ will begin to depend on Λ as well as on the suspension coordinate, and so the spectral lines will begin to broaden uniformly into bands.

In the case of Taylor flow the rotational symmetry of the apparatus will induce a double suspension $\Sigma^2 A \to \Sigma^2\Lambda$. Therefore the quasi-periodic torus T^2 explodes into a 3-dimensional strange attractor, accompanied by a broadening of the quasi-periodic spectral lines.

8.5. Summary. Strange bifurcations involving Ω-explosions from periodic attractors to strange attractors have properties that resemble the onset of turbulence, as follows:

(i) catastrophic qualitative change;
(ii) continuous quantitative change;
(iii) expanding properties, implying sensitive dependence on initial condition, and chaotic appearance;

(iv) indecomposable properties, implying that time-average measurements are independent of initial condition;
(v) mixing properties, implying the uniform broadening of frequency spectral lines into bands;
(vi) stability properties, implying the stability of turbulent flow, and the repeatability of measurements.

It should be emphasised, however, that much of the mathematics is incomplete as yet, and still developing.

References

1. R. Abraham and J. E. Marsden, *Foundations of mechanics* (2nd edition), Benjamin, Reading, MA 1978.
2. R. Abraham and S. Smale, Nongenericity of Ω-stability, *Global Analysis, Proc. Symp. Pure Math. Vol. 14*, Am. Math. Soc., Providence, RI, 1970, 5–8.
3. A. Andronov and L. Pontryagin, Systèmes grossiers, *C.R.* (*Dokl.*) *Acad. URSS* **14** (1937), 247–251.
4. D. V. Anosov, Geodesic flows on closed Riemannian manifolds with negative curvature, *Proc. Steklov Inst. Math.* **90** (1967), 1–235.
5. V. I. Arnold, Criticial points of smooth functions and their normal forms, *Russian Math. Surveys* **30** (1975), 1–75.
6. T. B. Benjamin, Bifurcation phenomena in steady flows of a viscous fluid, in *Proc. Roy. Soc. Lond. A* **359** (1978), 1–43.
7. P. Bernard and T. Ratiu (Eds.), *Turbulence Seminar*, Lecture Notes in Math. Vol. 615, Springer-Verlag, Berlin, 1977.
8. R. Bowen, A model for Couette flow data, *Turbulence Seminar*, Lecture Notes in Math. Vo. 615, Springer-Verlag, Berlin, 1977, 117–133.
9. R. Bowen and D. Ruelle, The ergodic theory of Axiom *A* flows, *Inventiones Math.* **29** (1975), 181–202.
10. J. J. Callahan, Special bifurcations of the double cusp, Preprint, Smith College, Northampton, MA 1978.
11. M. M. Couette, Études sur le frottement des liquides, *Ann. Chim. Phys.* **6**, Ser. 21 (1890), 433–510.
12. P. R. Fenstermacher, H. L. Swinney, and J. P. Gollub, Dynamical instabilities and the transition to chaotic Taylor vortex flow, *J. Fluid Mech.* **94** (1979), 103–128.
13. J. Gollub and H. L. Swinney, Onset of turbulence in a rotating fluid, *Phys. Rev. Lett.* **35** (1975), 927–930.
14. J. P. Gollub and S. V. Benson, Many routes to turbulent convection, *J. Fluid Mech.* **100** (1980), 449–470.
15. M. Golubitsky, An introduction to catastrophe theory and its applications, *SIAM Rev.* **20** (1978), 352–387.
16. M. Golubitsky and D. Schaeffer, A theory for imperfect bifurcation via singularity theory, *Commun. Pure and Appl. Math.* **32** (1979), 21–98.
17. M. Golubitsky and D. Schaeffer, Imperfect bifurcation in the presence of symmetry, *Commun. Math. Phys.* **67** (1979), 205–232.
18. M. Gorman, H. L. Swinney, and D. Rand, Quasi-periodic circular Couette flow: Experiments and predictions from dynamics and symmetry, submitted to *Phys. Rev. Lett.*, 1980.

19. J. Guckenheimer, Bifurcation and catastrophe, in *Dynamical Systems*, M. M. Peixoto, Ed., Academic, New York, NY, 1973, 95–110.
20. J. Guckenheimer, A strange strange attractor, *The Hopf Bifurcation*, Marsden and McCracken, Eds., Appl. Math. Series 19, Springer-Verlag, Berlin, 1976, 368–381.
21. W. Güttinger and H. Eikemeier (Eds.), *Structural Stability in Physics*, Synergetics Vol. 4, Springer-Verlag, Berlin, 1979.
22. J. Hayden and E. C. Zeeman, 1980 Bibiliography on catastrophe theory, Univ. Warwick, Coventry.
23. M. Hénon, A two-dimensional mapping with a strange attractor, *Commun. Math. Phys.* **50** (1976), 69–77.
24. P. J. Holmes, A non-linear oscillator with a strange attractor, *Phil. Trans. Roy. Soc.* **292** (1979), 419–448.
25. E. Hopf, Abzweigung einer periodischen Lösung von einer stationären Lösung einer Differentialsystems, *Ber. Verh. Sachs. Akad. Wiss. Leipzig Math. Phys.* **95** (1943), 3–22.
26. L. D. Landau and E. M. Lifshitz, *Fluid Mechanics*, Pergamon, Oxford, 1959.
27. E. Looijenga, On the semi-universal deformation of a simple-elliptic singularity, Part I: Unimodularity, *Topology*, **16** (1977), 257–262.
28. E. N. Lorenz, Deterministic nonperiodic flow, *J. Atmos. Sci.* **20** (1963), 130–141.
29. R. Lozi, Un attracteur étrange (?) du type attracteur de Hénon, *J. Phys.* **39**, C5 (1978), 9–10.
30. L. Markus, *Lectures in Differentiable Dynamics*, CBMS Regional Conf. Ser. in Math. 3, Amer. Math. Soc., Providence, RI, 1971; revised 1980.
31. L. Markus, Extension and interpolation of catastrophes, *Ann. NY Acad. Sci.* **316** (1979), 134–149.
32. J. E. Marsden and M. McCraken, *The Hopf Bifurcation and Its Applications*, Applied Math. Series 19, Springer-Verlag, Berlin, 1976.
33. J. Mather, Stability of C^∞-mappings, *Publ. Math. IHES* **35** (1968), 127–156 and **37** (1969), 223–248.
34. M. Misiurewicz, Strange attractors for Lozi mappings, in *Proc. Symp. Dynamical Systems & Turbulence*, Warwick 1980, Lecture Notes in Math., Springer-Verlag, Berlin (to appear).
35. A. S. Monin, On the nature of turbulence, *Sov. Phys. Usp.* **21**, 5 (1978), 429–442.
36. S. Newhouse and J. Palis, Bifurcations of Morse–Smale dynamical systems, in *Dynamical systems*, M. M. Peixoto, Ed., Academic, New York, NY, 1973, 303–366.
37. S. Newhouse, Diffeomorphisms with infinitely many sinks, *Topology* **13** (1974), 9–18.
38. S. Newhouse, D. Ruelle, and F. Takens, Occurrence of strange Axiom A attractors near quasi periodic flows on T^m, $m \geq 3$, *Commun. Math. Phys.* **64** (1978), 35–40.
39. J. Palis and S. Smale, Structural stability theorems, *Global Analysis*, Proc. Symp. Pure Math. 14, Am. Math. Soc., Providence, RI, 1970, 223–231.
40. W. Parry, Cocycles and velocity changes, *J. Lond. Math. Soc.* (2) 5 (1972) 511–516.
41. Ja. B. Pesin, Lyapunov characteristic exponents and smooth ergodic theory. *Russian Math. Surveys* **32**, 4 (1977), 55–114.
42. R. V. Plykin, Sources and currents of A-diffeomorphisms of surfaces, *Math. Sb.* **94**, 2(b) (1974), 243–264.
43. T. Poston and I. N. Stewart, *Catastrophe Theory and Its Applications*, Pitman, London, 1978.
44. C. C. Pugh and M. Shub, The Ω-stability theorem for flows, *Invent. Math.* **11** (1970), 150–158.

45. D. Rand, The topological classification of Lorenz attractors, in *Proc. Camb. Phil. Soc.* **83** (1978), 451–460.
46. —, Dynamics and symmetry: Predictions for modulated waves in rotating fluids, *Arch. Rat. Mech.*, to appear.
47. D. Rand and E. C. Zeeman, Models of the transition to turbulence in certain hydrodynamical experiments, in *Proc. Symp. Dynamical Systems & Turbulence, Warwick 1980*, Lecture Notes in Maths, Springer-Verlag, Berlin, to appear.
48. J. Robbin, A structural stability theorem, *Ann. Math.* **94** (1971), 447–493.
49. R. C. Robinson, Structural stability of vector fields, *Ann. Math.* **99** (1974), 154–175.
50. O. E. Rössler, Chaos, *Structural Stability in Physics*, W. Güttinger and H. Eikemeier, Eds., Synergetics Vol. 4, Springer-Verlag, Berlin, 1979, 290–309.
51. D. Ruelle and F. Takens, On the nature of turbulence, *Comm. Math. Phys.* **20** (1971), 167–192 and **23** (1971), 343–344.
52. D. Ruelle, A measure associated with Axiom *A* attractors, *Amer. J. Math.* **98** (1976), 619–654.
53. —, The Lorenz attractor and the problem of turbulence, in *Quantum Dynamics Models and Mathematics*, Lecture Notes in Math. Vol. 565, Springer-Verlag, Berlin, 1976, 146–158.
54. —, Sensitive dependence on initial condition and turbulent behaviour of dynamical systems, *Ann. NY Acad. Sci.* **316** (1978), 408–416.
55. —, Les attracteurs étranges, *La Recherche* **108**, 11 (1980), 132–144.
56. —, Ergodic theory of differentiable dynamical systems, *Publ. Math. IHES* **50**, 1979, 27–58.
57. —, On the measures which describe turbulence, preprint, IHES, 1978.
58. D. Schaeffer, Qualitative analysis of a model for boundary effects in the Taylor Problem, *Math. Proc. Camb. Phil. Soc.* **87** (1980), 307–337.
59. F. J. Seif, Cusp bifurcation in pituitary thyrotropin secretion, in *Structural Stability in Physics*, W. Güttinger and H. Eikermeier, Eds., Synergetics Vol. 4, Springer-Verlag, Berlin, 1979, 275–289.
60. R. Shaw, Strange attractors, chaotic behaviour, and information flow, Preprint, Univ. Cal., Santa Cruz, 1978.
61. M. Shub, Thesis, Univ. California, Berkeley, 1967.
62. S. Smale, A structurally stable differentiable homeomorphism with an infinite number of periodic points, in *Proc. Int. Symp. Non-linear Vibrations, Vol. II*, 1961; *Izdat. Akad. Nauk Ukrain, Kiev* (1963), 365–366.
63. —, Diffeomorphisms with many periodic points, *Differential and Combinatorial Topology*, Princeton Univ. Press, Princeton, NJ, 1965, 63–80.
64. —, Differential dynamical systems, *Bull. Amer. Math. Soc.* **73** (1967), 747–817.
65. —, The Ω-stability theorem, *Global Analysis*, Proc. Symp. Pure Math. Vol. 14, Am. Math. Soc., Providence, RI, 1970, 289–298.
66. —, Dynamical systems and turbulence, in *Turbulence Seminar*, Lecture Notes in Math. Vol. 615, Springer-Verlag, Berlin, 1977, 48–70.
67. H. L. Swinney and J. P. Gollub, The transition to turbulence, *Physics Today*, **31**, 8 (1978), 41–49.
68. F. Takens, Unfoldings of certain singularities of vector fields: Generalised Hopf bifurcations, *J. Diff. Eqs.* **14** (1973), 476–493.
69. —, Singularities of vector fields, *Publ. Math. IHES* **43** (1974), 47–100.
70. —, Detecting strange attractors in turbulence, in *Proc. Symp. Dynamical Systems and Turbulence, Warwick 1980*, Lecture Notes in Math., Springer-Verlag, Berlin, to appear.
71. G. I. Taylor, Stability of a viscous liquid contained between two rotating cylinders, *Phil. Trans. Roy. Soc. A* **223** (1923), 289–343.

72. R. Thom, *Structural Stability and Morphogenesis* (Engl. trans. D. Fowler, 1975) Benjamin, New York, NY, 1972.
73. —, Structural stability, catastrophe theory and applied mathematics, *SIAM Review* **19** (1977), 189–201.
74. R. F. Williams, Expanding attractors, *Publ. Math. IHES* **43** (1974), 161–203.
75. —, The structure of Lorenz attractors, *Publ. Math. IHES*, **50** (1979), 73–100.
76. E. C. Zeeman, *Catastrophe Theory: Selected Papers 1972–1977*, Addison-Wesley, Reading, MA, 1977.

The Emphasis on Applied Mathematics Today and Its Implications for the Mathematics Curriculum

Peter J. Hilton*

1. Across the country, and beyond the borders of the United States, the cry is being heard that we mathematicians should be concerning ourselves more, both in our research and in our teaching, with applications of mathematics. It is being argued that we have been overemphasizing mathematics itself, the autonomous discipline of mathematics, at the expense of due attention to its usefulness, to its role in science, in engineering, in the conduct of modern society. Some put it crudely—there is too much "pure mathematics," too little "applied mathematics."

The argument is an important one; it is rendered the more crucial by its relation to a critical problem now confronting the profession of academic mathematicians—the declining enrollment in the mathematics major and, in particular, in the more traditional upper division courses. How can this process be arrested and reversed? The answer is seen to be closely connected with the idea that we should somehow seek to make our mathematics courses more relevant to the needs and interests of today's students, without any sacrifice of standards or of integrity.

Many groups of mathematicians and mathematics educators have devoted considerable effort to coming to grips with these related problems. There is the report† of the NRC Committee on Applied Mathematics Training; a panel of CUPM, under the chairmanship of Professor Alan Tucker, is in the process of producing sample curricula with a decidedly "applied" flavor; there is a joint MAA–SIAM Committee considering undergraduate and graduate courses; there are the recommendations of the PRIME 80 Conference; and much else.

* Department of Mathematics, Case Western Reserve University, Cleveland OH 44106.

† The Role of Applications in the Undergraduate Mathematics Curriculum, AMPS, National Research Council, Washington (1979).

At this conference we have listened to some of the best applied mathematicians describing recent work in their fields. With these stimulating talks fresh in our minds it is natural to ask ourselves, as teachers of mathematics, two questions. First, what emerges from these talks as the distinctive quality of good applied mathematics; and, second, how can we prepare our students to do the kind of work these mathematicians do and which they have so vividly described.

First it is plain, above all else, from these talks that to do good work in applied mathematics one needs to be a good mathematician. Of course, one needs more, but one certainly needs this. Moreover, there is no natural division of mathematics itself into applicable mathematics and mathematics *sui generis*—our nine speakers have used, in their talks, apart from the obvious areas of ordinary and partial differential equations, material from combinatorics, commutative algebra, the theory of jets, algebraic geometry, Lie groups and Lie algebras, differential topology, algebraic topology, fibre bundle theory, deformation of complex structures, singularity theory and functional analysis.* Moreover, the talks have been distinguished from talks in so-called pure mathematics not by the manner of treatment, or the rigor of the argument, but by the "real world" motivation for the mathematical problem. Thus if our students are to be able to follow in these footsteps, they must have a broad and deep education in mathematics and an attitude of a very positive kind towards applications. This conclusion is in striking agreement with that of the NRC Committee referred to above, where both points are made with great emphasis, and where, in keeping with the conclusion that there is a fundamental unity encompassing the whole of mathematics, pure and applied, a plea is made for a broad-based major in the mathematical sciences, giving *all* students the encouragement and opportunity to acquire familiarity with the way mathematics is, in fact, applied.

Can we then make a reasonable working definition of applied mathematics? Let me attempt one, based on a definition of applied analysis given by Kaper and Varga.† We propose the following:

> "The term *applied mathematics* refers to a collection of activities directed towards the formulation of mathematical models, the analysis of mathematical relations occurring in these models, and the interpretation of the analytical results in the framework of their intended application. The objective of an applied mathematics research activity is to obtain qualitative and quantitative information about exact or approximate solutions. The methods used are adapted from all areas of mathematics; because of the universality of mathematics, one analysis often leads, simultaneously, to applications in several diverse fields."

* It is true that the talk by Professor Oster is different in kind from the others. It was the deliberate plan of the program committee to put into the program one such exceptional talk. The plan was splendidly vindicated!

† Hans G. Kaper and Richard S. Varga, Program Directions for Applied Analysis, Applied Mathematical Sciences Division, Department of Energy (1980). I am most grateful for the opportunity to see this paper, from which I have drawn many ideas.

In this paper we will first make some general remarks about the nature of mathematical modeling—these will form the content of the next section. We will then suggest, in the third section, that the basic method of applied mathematics is not, in fact, as distinctive as it at first appears; that, in fact, it has much in common with processes of abstraction and generalization that go on within mathematics itself. These remarks will lead us to the conclusion that, by modifying our approach to the curriculum in certain ways that give expression to the unity of the mathematical sciences which it is our task to present to our students, we may prepare them to become mathematicians, able and willing to place their knowledge and talents at the service of problems coming from within or without mathematics. Thus might the sterile antagonism which one sometimes finds today between pure and applied mathematics—and pure and applied mathematicians—be eliminated by abandoning these labels and reverting to the notion of a single indivisible discipline, mathematics.

2. What can be said in general of the process of mathematical modeling? The following schema seems to reflect the methods described by the illustrious—and successful—mathematicians who have spoken at this conference:

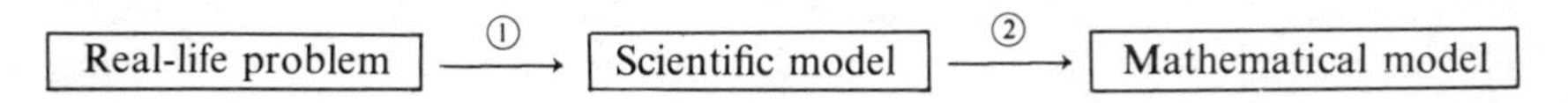

Step ① occurs whenever a mature science is involved; it may well happen in the soft social sciences that one proceeds directly from real-life problem to mathematical model. Such a process is dangerous, because it is within the scientific model that one locates the measurable constructs about which one theorizes, quantitatively and qualitatively, within the mathematical model. Thus the direct passage from problem to mathematical model, while often intellectually exciting, is open to the objection that one may well be using sophisticated mathematical tools to reason about extremely vague concepts involving very unreliable measurements.

Let us then assume that a scientific model is, indeed, articulated. This will consist, typically, of *objects* (observables, constructs) and *laws* (physical, chemical, biological) about the behavior of matter in the form of liquids, gases and solid particles. The selection of the appropriate scientific model, step ①, may be called *constructive analysis*; for example, in the study of energy systems (fission reactors, combustion chambers, coal gasification plants), the laws express the rate of flow of mass, momentum and energy between the components of the system.

Step ② consists of choosing a mathematical model for the analysis of the scientific system. The mathematical model is, by its very nature, both more abstract and more general than the scientific system being modeled. Thus the conservation laws for an incompressible viscous fluid lead to the Navier–Stokes equation, which is an *evolution* equation; but we may also derive *equi-*

librium equations leading to the study of bifurcation phenomena. Again, the study of non-linear wave mechanics may lead to the Korteweg–de Vries (KdV) equation for the behavior of long water waves in a rectangular channel.* Here the theory predicts solitary waves which interact but emerge unchanged—these are the solitons of modern theoretical physics. In these examples, the original constructive analysis leads to the next stage of mathematical analysis (in these examples, qualitative analysis) within the mathematical models. Typically, again, this stage consists of proofs of existence of solutions, together with a study of their uniqueness, stability (sensitivity to changes of parameter), and behavior over large time intervals (asymptotic analysis).

Quantitative analysis, however, also plays a key role in today's applied mathematics, due largely to the general availability of the high-speed computer. Numerical methods of a sophisticated kind have been developed; asymptotic and perturbation methods are widely used. One particularly important new tool of quantitative analysis that one may mention here is the "finite element" method, invented by Richard Courant, but rediscovered by engineers who saw its potential in conjunction with the computer. The method itself is, of course, undergoing improvement and refinement.

In several of the talks presented at this conference, we have seen evidence that our schema is often incomplete in one significant respect—the process of elaborating a mathematical model may well be *iterated*. Thus our scientific model may lead to a first-order differential equation which can be interpreted as a dynamical system and thus embedded in the theory of vector bundles or, more generally, fibre bundles. Thus the existence of a solution is translated, first, into an integrability problem and then into a cross-section problem to which we may apply the techniques of obstruction theory. Thus it would be very misleading to think of the process of abstraction and generalization as a one-stage procedure; by the same token it is a mistake to think of an area of mathematics as ineffably "pure" because all its direct, immediate contacts are with other areas of mathematics. This simplistic view would have found no favor with our speakers at this conference, whose views on the applicability of mathematics and on the relations between so-called pure and applied mathematics excluded any possibility of a rigid distinction being made.

Another very striking feature emerged from a consideration of the contributions to this conference. The mathematical content of the talks bore strong testimony to that universality of mathematics to which attention is drawn in the description of the nature of applied mathematics quoted above. For mathematicians adopting very different starting points in their investigations—control systems, the study of porous media, embryogenesis, bifurcation theory and turbulence—found themselves concerned with

* More precisely, the KdV equation describes the propagation of waves of small amplitude in a dispersive medium.

significantly overlapping domains of mathematics in the design and analysis of their models. The Navier–Stokes equation, for example, together with its linearization, figured in many contributions; and questions of stability, naturally, arose frequently when systems of partial differential equations were involved. Thus it may fairly be said that the talks displayed the salient features of good mathematics in action—its unity, its subtlety, its diversity, its power and its universality.

It is plain that if our students are to be able to apply mathematics effectively, they must gain some understanding and experience of the art of mathematical modeling. We do not recommend a special modeling course; rather, the modeling process should be explicitly discussed when an application of mathematics to a scientific problem is in question. We would ourselves recommend that the discussion include the following component items: the selection of a suitable problem; the development of an appropriate model; the collection of data; reasoning within the model (qualitative analysis); calculations (quantitative analysis); reference back to the original problem to test the validity of a solution; modification of the model; generalization of the model as a conceptual device. Moreover, the entire modeling process must be set against the contemporary background of a strong computer capability (actual or assumed).

However, it will be our claim in the next section that it is unnecessary to separate applications from the rest of mathematical activity in order to emphasize the processes named above. We will suggest, in fact, that good "applied" mathematics and good "pure" mathematics have a great deal in common, and that this complementarity should be reflected in the undergraduate curriculum.

3. The question I wish to consider in this section is this—how special to applied mathematics are the techniques and procedures described in the previous section? Of course, in one sense they certainly are special; for if we start off with a "real world" problem and apply mathematical reasoning, we are *ipso facto* doing applied mathematics. Thus for the question to make any sense, we must allow that the original problem to be tackled mathematically could itself be a mathematical problem. We would then claim that the process of abstraction which is characteristic of the schema we described in Section 2 also features in work within mathematics itself. Let us immediately give an example; this is admittedly a relatively trivial piece of mathematics, but it is without doubt a piece of mathematics currently pertaining to the undergraduate curriculum. We allow ourselves, here and subsequently, to consider the closely related processes of abstraction and generalization.

Suppose that it is observed that $5^6 \equiv 1 \bmod 7$. This may be verified empirically—thus we compute $5^6 - 1$, obtaining 15624 and check that 15624 is exactly divisible by 7. This argument is compelling and convincing—but unsatisfactory. We do not feel, with this demonstration, that we understand *why* the assertion is true.* The situation is ripe for generalization; we make a mathematical model! We conjecture, and then prove, that if p is any prime number and if a is prime to p, then (Fermat's Theorem) $a^{p-1} \equiv 1 \bmod p$. Notice that this is a generalization, not an abstraction, because we are still talking about rational integers. However, we may not feel entirely satisfied with the generality of Fermat's Theorem. We could proceed in one direction to Euler's Theorem; or we could regard Fermat's Theorem as itself a special case of Lagrange's Theorem that the order of a subgroup of a finite group divides the order of the group. This latter development involves abstraction as well as generalization, for we are now discussing abstract groups, and postulate, in our abstract system, only one binary operation (whereas the integers admit two, and both were involved in our original demonstration that $5^6 \equiv 1 \bmod 7$), which need not even be commutative.

A particular feature of this example is the iteration of the modeling (generalizing) process; as we saw, this is also, frequently, a feature of applied mathematics. On the other hand, let us immediately admit that there is also a difference between modeling a non-mathematical problem, and the modeling we did here. In our example, we obtained incontrovertible proofs of our original congruence, and, of course, of related congruences, whereas, in applied mathematics, the mathematical reasoning can at best establish a scientific assertion as a good working hypothesis, a good approximation to the truth. Let us, however, give a second "mathematical" example to show that this difference is not so absolute.

There is a beautiful numerical process, based on our base 10 enumeration system, called *casting out* 9's. What is involved here, in mathematical terms, is the canonical ring projection $\theta\colon \mathbb{Z} \to \mathbb{Z}/9$; we use the residue ring $\mathbb{Z}/9$ because it is particularly easy to compute† θ. Now we may say that θ provides a means of modeling identities in $\mathbb{Z}$ by means of identities in $\mathbb{Z}/9$. This is a good checking procedure, because it is far easier to do calculations in $\mathbb{Z}/9$ than in $\mathbb{Z}$. However, we cannot *prove* identities in $\mathbb{Z}$ by modeling them by true identities in $\mathbb{Z}/9$; we can only *disprove* them by modeling them by false identities. We are here involved in the important mathematical process of *simplification* (with preservation of structure); there is a strong analogy here with the simplification involved in modeling a real-world situation.

If we are to do justice to abstraction, generalization and simplification as key processes within mathematics itself, we find ourselves led inevitably to give prominence to the essential unity of mathematics. In practical terms this means we must insist far less on the autonomy and (apparent) independence of the various mathematical disciplines and emphasize their (real) inter-

* We are thus in the unfortunate situation so typical of our students! They are compelled to accept but do not truly understand.

† In fact, θ extends, uniquely, to a ring homomorphism $\mathbb{Z}_3 \to \mathbb{Z}/9$, where $\mathbb{Z}_3$ is the ring of integers localized at the prime 3. Likewise the projection $\mathbb{Z} \to \mathbb{Z}/11$ is easily computable and extends to $\mathbb{Z}_{11} \to \mathbb{Z}/11$.

dependence. This poses severe problems in the design of curricula, but we believe that perhaps the most important desideratum is the breadth of view of the instructor.

Examples of interaction between different mathematical disciplines abound, at the undergraduate as at any other level. Thus we use topology in the foundations of real analysis (a continuous function from a compact metric space to a Hausdorff space is uniformly continuous); we use algebra to do topology (the fundamental group); we use complex variable theory to do algebra (the fundamental theorem of algebra); we use algebra to do geometry (the syzygy method for proving Desargues' Theorem in the coordinatized plane); we use linear algebra to study systems of linear differential equations (eigenvalues and eigenvectors). These and other examples can be presented as modeling one mathematical situation by means of another. It is our contention that this "applied perspective" should be adopted in mathematics itself—and not merely for the worthy reason that this will help the student to become familiar with the ways of doing applied mathematics!*

Are there essential differences between the methodologies of "pure" and "applied" mathematics? This is, in my view, a very interesting subject for research, with strong implications for the teaching of mathematics. My own thinking is still at a fairly primitive stage on this question, but let me offer one fairly obvious example of an essential difference.

* The place of geometry in the curriculum is, today, a special concern and a special problem. Students are arriving at our universities and colleges woefully ignorant of geometry and seriously lacking in any geometric intuition. These failings undoubtedly contribute to the difficulties they experience with the regular calculus sequence. Among the upper division courses we also find that courses in geometry are under-subscribed (along with certain other "traditional" offerings). Indeed it may happen that the only viable geometry course is a course designed for future high school teachers—this is viable in the sense that it will have an adequate enrollment, but it usually fails to do justice to geometry as a living branch of mathematics.

We would recommend that the geometric point of view figure prominently in virtually all undergraduate mathematics courses. This point of view allows one to conceptualize more easily; and geometry is a wonderful source of ideas and questions. Geometry, in this informal sense, may be thought of as partaking of the quality of both pure and applied mathematics—it is, after all, of all the branches of mathematics, that which is closest to the world of our experience.

It is probably not realistic to recommend an attempt to revive the study of geometry for its own sake, in courses devoted exclusively to the discipline. But geometry is very "real" to the students; it provides questions to which the disciplines of analysis and algebra provide answers. Without geometry, these latter disciplines must often seem to the students to answer questions they could never imagine themselves asking!

I would like to join Bert Kostant in making a special plea for a regular course in the curriculum on Lie groups. Here geometry, algebra and analysis come together in a theory of great power and importance to both pure and applied mathematics; it is, moreover, a subject rich in history.

Naturally, at the graduate level, we should find a great interest among faculty and students in algebraic and differential geometry, in view of the very significant advances currently being made in these subjects, as autonomous disciplines, in their relations to other parts of mathematics, and as suitable models for problems coming from engineering and the physical sciences.

Suppose we are modeling some physical phenomenon and produce a differential equation with certain boundary conditions. Suppose further that we prove that this mathematical system has no solution. The effect of this discovery is to discredit the model—we must have over-simplified (say, by linearizing) or we must have neglected some aspect of the physical situation which was, in fact, highly relevant to the dynamic process we were modeling. However, if we model a mathematical situation, the connection between the situation and the model is far closer. There is still, as in applied problems, the very difficult art of choosing a good model (e.g., a useful generalization); but if the problem in the model has no solution, then the original problem had no solution, either. This remains true whether we are generalizing or simplifying in constructing our model.

4. We would like to close this essay with a few remarks on problem-solving as a curricular or pedagogic device. The clamor for applications finds its echo at the pre-college and even undergraduate level in a strong plea (endorsed by the PRIME 80 Conference and the National Council of Teachers of Mathematics, as a key element in their platforms) for greater emphasis on problem-solving.

Now it is not in question—and should always have been obvious—that the principal reason for learning mathematics is that it enables one to solve problems. If certain programs have appeared to neglect this proposition then they are undoubtedly, to that extent, seriously defective. However, what is emerging from all the propaganda for problem-solving tends to be something very different in nature from a simple forceful recommendation to keep in mind why we learn mathematics. For the advocates of problem-solving seem to be arguing that we should be teaching problem-solving as an *alternative* to our traditional approach. Good pedagogical strategy should be "problem-oriented," they argue; and, if problem-solving is effectively taught, we need not trouble the students so much to absorb the "theory" which has hitherto proved a stumbling-block to them. An example of this attitude is to be found in the publishers' puff for a (very good) book on applied combinatorics which reads, "Its applied approach gives your students the *emphasis on problem-solving* that they need to participate in today's new fields. Rather than focus on theory, this text contains hundreds of worked examples with discussion of common problem-solving errors . . ."

Not for the first time, I must insist that this false dichotomy between the building and analysis of mathematical structure on the one hand and problem-solving on the other is dangerous. If problems are to be solved mathematically, the mathematical model must be chosen. Either it must already be available to the would-be problem-solver or he (or she) must be capable of developing it by the modification of existing mathematical models. Thus the investigator absolutely needs to know, to understand, and to be able to discriminate between different mathematical structures.

There are, it is true, certain problem-solving strategies and precepts which it is worthwhile enunciating explicitly. But these cannot serve as a substitute for a knowledge of mathematics. One of the most impressive features of the talks at this conference was the evidence of the vast array of mathematical knowledge at the disposal of the speakers, and of how much they actually needed for the particular investigations they described. It would be truly calamitous if the belief were to be spread abroad that difficult quantitative problems could be solved merely by becoming proficient in the field of problem-solving, allying a knowledge of a few general principles to "sound common sense." This is just a sophisticated restatement of the old egalitarian fallacy.

There is a further reason, too, less obvious at first, why it would be a mistake to concentrate so exclusively on problem-solving. For it is implicit in the concept of problem-solving that the problem has already been formulated; it is further suggested that it is a matter then of selecting the best mathematical model and successfully exploiting it. Thus the question of *how* we formulate good questions is totally ignored—and this is an essential question in scientific work. Moreover, the problem-solving approach ignores the fact that it is often the mathematical concept and the mathematical result which suggest the promising question. Frequently it is an advance in mathematics which enables us to see the true nature of a scientific problem more clearly and to pose the significant questions (this is eminently true, for example, of electromagnetic theory and, more recently, the theory of solitons). Thus the essential two-way flow between mathematics and science is lost in an exclusively problem-solving mode of instruction.*

None of this, of course, is to gainsay the stimulus which the attempt to solve interesting problems provides for the understanding and doing of mathematics. The UMAP modules can be extremely valuable as a component of a rich mathematical education. It should not be thought, however—and here I believe I have understood the intentions of the UMAP editorial board—that these modules are to be concentrated on applied topics. It is easy to supply a list of mathematical topics suitable for modular treatment; a brief sample might include maximum and minimum problems treated by various methods; thought-provoking paradoxes; linearization and linearity in mathematics; computational complexity; the geometry of 3-dimensional polyhedra; topics in combinatorics; algebraic curves; the classical groups. However, as these topics should suggest, the modules do not replace the systematic study of mathematics—they stimulate, enrich and enliven it. Ultimately we can only serve the purpose of mathematics education, in all its facets, by inculcating both the ability and the will to do mathematics and to use it. If, as some enthusiasts for a more applied curriculum and for more problem-solving rightly claim, the ability without the will leads to sterility, it is also true that the will without the ability leads to frustration. We avoid both these unpleasant consequences by teaching all of mathematics as a unity, emphasizing its unique generality and its immense power.

* Problem-solving may be characterized as "going from question to answer;" but scientific and mathematical progress often consists of going from answer to question.